D1758674

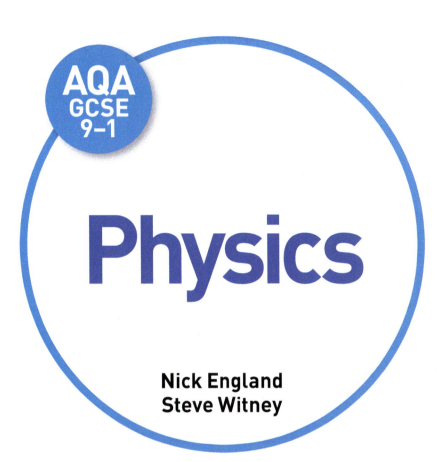

AQA GCSE 9-1

Physics

Nick England
Steve Witney

Approval message from AQA

This textbook has been approved by AQA for use with our qualification. This means that we have checked that it broadly covers the specification and we are satisfied with the overall quality. Full details of our approval process can be found on our website.

We approve textbooks because we know how important it is for teachers and students to have the right resources to support their teaching and learning. However, the publisher is ultimately responsible for the editorial control and quality of this book.

Please note that when teaching the *AQA GCSE Physics* course, you must refer to AQA's specification as your definitive source of information. While this book has been written to match the specification, it cannot provide complete coverage of every aspect of the course.

A wide range of other useful resources can be found on the relevant subject pages of our website: www.aqa.org.uk.

HODDER EDUCATION
AN HACHETTE UK COMPANY

Although every effort has been made to ensure that website addresses are correct at time of going to press, Hodder Education cannot be held responsible for the content of any website mentioned in this book. It is sometimes possible to find a relocated web page by typing in the address of the home page for a website in the URL window of your browser.

Hachette UK's policy is to use papers that are natural, renewable and recyclable products and made from wood grown in well-managed forests and other controlled sources. The logging and manufacturing processes are expected to conform to the environmental regulations of the country of origin.

Orders: please contact Hachette UK Distribution, Hely Hutchinson Centre, Milton Road, Didcot, Oxfordshire, OX11 7HH. Telephone: +44 (0)1235 827827. Email education@hachette.co.uk Lines are open from 9 a.m. to 5 p.m., Monday to Friday. You can also order through our website: www.hoddereducation.co.uk

© Nick England and Steve Witney 2016

First published in 2016 by
Hodder Education,
An Hachette UK Company
Carmelite House
50 Victoria Embankment
London EC4Y 0DZ

Impression number   8

Year   2022

All rights reserved. Apart from any use permitted under UK copyright law, no part of this publication may be reproduced or transmitted in any form or by any means, electronic or mechanical, including photocopying and recording, or held within any information storage and retrieval system, without permission in writing from the publisher or under licence from the Copyright Licensing Agency Limited. Further details of such licences (for reprographic reproduction) may be obtained from the Copyright Licensing Agency Limited, www.cla.co.uk

Cover photo mark penny – Fotolia
Illustrations by Aptara

Typeset in OfficinaSans-Book, 11.5/13 pts by Aptara Inc.
Printed and bound by CPI Group (UK) Ltd, Croydon, CR0 4YY

A catalogue record for this title is available from the British Library.

ISBN 9781471851377

Contents

Get the most from this book

Welcome to the AQA GCSE Physics Student Book.

This book covers the Foundation and Higher-tier content for the 2016 AQA GCSE Physics specification.

The following features have been included to help you get the most from this book.

Prior knowledge

This is a short list of topics you should be familiar with before starting a chapter. The questions will help to test your understanding. Extra help and practice questions can be found online in our AQA GCSE Science Teaching & Learning Resources

KEY TERMS

Important words and concepts are highlighted in the text and clearly explained for you in the margin.

Practical

These practical-based activities will help consolidate your learning and test your practical skills.

Required practical

AQA's required practicals are clearly highlighted.

TIPS

These highlight important facts, common misconceptions and signpost you towards other relevant chapters. They also offer useful ideas for remembering difficult topics.

Higher-tier only

Some material in this book is only required for students taking the Higher-tier examination. This content is clearly marked with the blue symbol seen here.

Examples

Examples of questions and calculations that feature full workings and sample answers.

Show you can...

Complete the Show you can tasks to prove that you are confident in your understanding of each topic.

Test yourself questions

These short questions, found throughout each chapter, allow you to check your understanding as you progress through a topic.

Chapter review questions

These questions will test your understanding of the whole chapter. They are colour coded to show the level of difficulty and also include questions to test your maths and practical skills.

■ Simple questions that everyone should be able to answer without difficulty.

■ These are questions that all competent students should be able to handle.

■ More demanding questions for the most able students.

Answers

Answers for all questions and activities in this book can be found online at:

www.hoddereducation.co.uk/aqagcsephysics

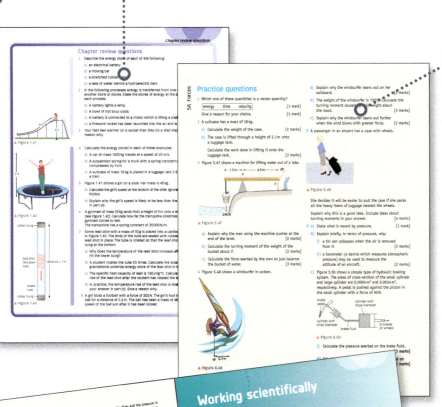

Practice questions

You will find Practice questions at the end of every chapter. These follow the style of the different types of questions you might see in your examination and have marks allocated to each question part.

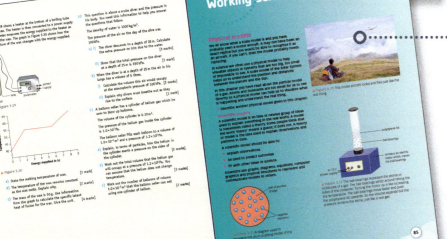

Working scientifically

In this book, Working Scientifically skills are explored in detail in the activity at the end of each chapter. Work through these activities on your own or in groups. You will develop skills such as Dealing with data, Scientific thinking and Experimental skills.

* AQA only approve the Student Book and Student eTextbook. The other resources referenced here have not been entered into the AQA approval process.

The Publisher would like to thank the following for permission to reproduce copyright material:

Photo credits

p. 1 © zentilia – Fotolia; **p. 2** © U. Gernhoefer – Fotolia; **p. 10** © Mikael Damkier – Fotolia; **p. 16** Nick England; **p. 18** *t* © PHB.cz – Fotolia; *b* © WILLIAM WEST/AFP/Getty Images; **p. 22** © Alex White – Fotolia; **p. 24** *t* Vladislav Gajic – Fotolia; *b* © Gary Eastwood Photography / Alamy Stock Photo **p. 25** *t* © ADRIANO PECCHIO – iStock – Thinkstock; *c* © Laurence Gough – Fotolia; *b* © Saskia Massink – Fotolia; **p. 36** © SeanPavonePhoto – Fotolia; **p. 41** © tigger11th – Fotolia; **p. 50** © SeanPavonePhoto – Fotolia; **p. 51** © Fatbob – Fotolia; **p. 52** © Jaroslaw Grudzinski – Fotolia; **p. 54** © TebNad – Fotolia; **p. 57** © sytilin – Fotolia; **p. 58** Hank Morgan/SPL; **p. 66** © Andrew Dunn / Alamy Stock Photo; **p. 85** © Ljupco Smokovski – 123RF; **p. 87** © Photography by Courtesy Decoding Science, Bond Life Sciences Center ; **p. 98** © Nailia Schwarz – 123RF; **p. 100** © AfriPics.com / Alamy Stock Photo; **p. 102** © Aflo/ REX Shutterstock; **p. 104** © MARTYN F. CHILLMAID/SCIENCE PHOTO LIBRARY; **p. 105** © Department of Nuclear Medicine and PETCT, Amrita Institute of Medical Sciences, Kochi, Kerala, India (http:// www.ijnm.in/article.asp?issn=0972-3919 year=2013 volume=28 issue=3 spage=152 epage=162 aulast=Gandhi); **p. 114** *t* © Rick Hyman – iStock – Thinkstock; *b* © sumnersgraphicsinc – Fotolia; **p. 115** © Monkey Business – Fotolia; **p. 116** © mikhail mandrygin – 123RF; **p. 119** *t* © Daniel Vorley/ LatinContent/Getty Images; *c* © Toutenphoton – Fotolia; **p. 124** © kris Mercer / Alamy Stock Photo; **p. 129** *t* © vladru – Thinkstock; *b* © Bartłomiej Szewczyk – Fotolia; **p. 136** © frantisek hojdysz – Fotolia; **p. 139** © popov48 – Fotolia; **p. 140** © Matt Tilghman – 123RF; **p. 146** © Christopher Ison / Alamy Stock Photo; **p. 148** © OLIVIER MORIN/AFP/Getty Images; **p. 152** c © jpmatz – Fotolia; **p. 154** © Gustoimages/Getty Images; **p. 157** © Mariusz Blach – Fotolia; **p. 158** © NASA; **p. 164** Open Government License; **p. 166** © bytesurfer – Fotolia; **p. 168** © Volvo Car UK Ltd; **p. 177** Peter Ginter / Science Faction / SuperStock ; **p. 180** © Martin Dohrn / Science Photo Library; **p. 181** © Martin Dohrn / Science Photo Library; **p. 182** © art_zzz – Fotolia; **p. 188** © Digital Vision/ Photolibrary Group Ltd/Wild at Work DV509; **p. 190** © jingjofire – iStock – Thinkstock; **p. 191** © Isabelle Limbach/ iStockphoto/Thinkstock; **p. 195** *t* © alexsalcedo – Fotolia; *b* © Jose Manuel Gelpi – Fotolia; **p. 200** *t* © Cultura RM Exclusive/Joseph Giacomin – Getty Images; *b* © PAUL RAPSON/SCIENCE PHOTO LIBRARY; **p. 201** © FRANCK FIFE/AFP/Getty Images; **p. 201** © gloszilla – Fotolia; **p. 201** © nevodka – Thinkstock; **p. 208** © Vasyl Yakobchuk/Hemera/Thinkstock; **p. 210** © Voyagerix – Fotolia; **p. 215** © dule964 – Fotolia; **p. 220** © CERN; **p. 222** © AlexStar – iStock via Thinkstock – Getty Images; **p. 224** © sciencephotos / Alamy; **p. 225** © SSPL/Getty Images; **p. 228** © Andrew Lambert Photography / Science Photo Library; **p. 236** © hxdyl – iStock via Thinkstock – Getty Images; **p. 245** © Cordelia Molloy/SPL ; **p. 246** © RAUL ARBOLEDA/AFP/Getty Images; **p. 247** © NASA-JHUAPL-SWRI; **p. 248** *l* © NASA; *r* © NASA; **p. 249** © Galaxy Picture Library / Alamy Stock Photo; **p. 250** *t* © European Space Agency; *c* © NASA; *b* © standret – iStock – Thinkstock; **p. 251** *l* © European Southern Observatory (ESO); *c* © NASA; *r* © NASA; **p. 253** © NASA, ESA, J. Hester, A. Loll (ASU)

t = top. *b* = bottom, *l* = left, *c* = centre, *r* = right

Every effort has been made to trace all copyright holders, but if any have been inadvertently overlooked, the Publisher will be pleased to make the necessary arrangements at the earliest opportunity.

1 Energy

We are concerned that energy reserves are running out. What does the future hold? How will we generate electricity for the projected World population of 10 billion people in 2050?

Specification coverage

This chapter covers specification points: 4.1.1 Energy changes in a system, and the ways energy is stored before and after such changes, 4.1.2 Conservation and dissipation of energy and 4.1.3 National and global energy resources.

Previously you could have learned:

- Our primary source of energy is the Sun.
- Food provides us with energy to live.
- Energy reaches us from the Sun in the form of electromagnetic radiation.
- Fossil fuels are a source of energy.
- We generate electricity using fossil fuels and other sources of energy.
- Energy can be transferred from a hot object to colder objects.
- Metals are good thermal conductors. They allow energy to be transferred quickly.
- Fluids (liquids and gases) transfer energy by convection.
- Energy is measured in joules, J.

Test yourself on prior knowledge

1 State three examples of how you use energy every day.
2 Give an example of a fossil fuel.
3 Why are metals good thermal conductors?

Energy changes in a system, and the ways energy is stored before and after such changes

○ Energy stores and systems

We can begin to understand energy by studying changes in the way energy is stored when a **system** changes. A 'system' is an object or a group of objects that interact. Here are some situations with which you should be familiar.

- **Throwing an object upwards**
  When you throw a ball upwards, just after the ball leaves your hand it has a store of kinetic energy. When the ball reaches its highest point, it has a store of gravitational potential energy. Just before you catch it again, it has a store of kinetic energy.

- **Boiling water in a kettle**
  When you turn on your electric kettle, the water in the kettle gets hotter. There is now more internal (or thermal) energy stored in the hot water than there was in the cold water.

- **Burning coal**
  When we burn coal there is a chemical reaction. Coal has a store of chemical energy which is transferred to thermal energy as it burns. A coal fire can warm up a room.

▲ **Figure 1.1** One store of energy is transferred to another store.

- **A car using its brakes to slow down**
  A moving car has a store of kinetic energy. When the car slows to a halt, it has lost this store of kinetic energy. The brakes exert a frictional force on the wheels, and the brakes get hot. The store of kinetic energy in the car has been transferred to a store of thermal energy in the brakes. This energy is then transferred to the surroundings.

- **Dropping an object which does not bounce**
  Just before the object hits the ground, it has a store of kinetic energy. After the object has stopped moving, the kinetic energy has been transferred to a store of internal energy in the object and the surroundings. So the object and the surroundings warm up a little. (You might hear a noise, but the energy carried by the sound is also transferred to the internal energy of the surroundings.)

- **Accelerating a ball with a constant force**
  We have a store of chemical potential energy in our muscles. When we throw a ball, our store of chemical potential energy decreases, and the ball's store of kinetic energy increases. The hand applies a force to the ball and does work to accelerate it.

- **Holding two magnets with north poles facing**
  When you hold two magnets with like poles facing, you can feel a force which repels the magnets from each other. When the magnets are close together there is a store of magnetic potential energy. When you release the magnets, they move apart. The magnets' store of magnetic potential energy has reduced and their store of kinetic energy has increased.

Energy stores

In the simple everyday events and processes that were described above, we identified objects that had gained or lost energy. For example, objects slow down or get hotter. We saw that the way energy is stored changes.

We use the following labels to describe the stores of energy you will meet:

- kinetic
- chemical
- internal (or thermal)
- gravitational potential
- magnetic
- electrostatic
- elastic potential
- nuclear.

Counting the energy

Energy is a quantity that is measured in joules, J. Large quantities of energy are measured in kilojoules, kJ, and megajoules, MJ.

$$1\,\text{kJ} = 1000\,\text{J}\ (10^3\,\text{J}) \qquad\qquad 1\,\text{MJ} = 1\,000\,000\,\text{J}\ (10^6\,\text{J})$$

The reason that energy is so important to us is that there is always the same energy at the end of a process as there was at the beginning. If we add up the total energy in all the stores, that number stays the same.

Figures 1.2, 1.3 and 1.4 show some examples of counting the energy.

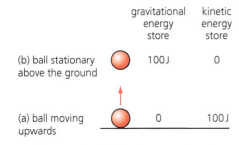

	gravitational energy store	kinetic energy store
(b) ball stationary above the ground	100 J	0
(a) ball moving upwards	0	100 J

▲ **Figure 1.2** A ball is thrown upwards from the ground. In (a) the ball has 100 J of kinetic energy and zero gravitational potential energy. In (b) the ball has 100 J of gravitational potential energy and zero kinetic energy.

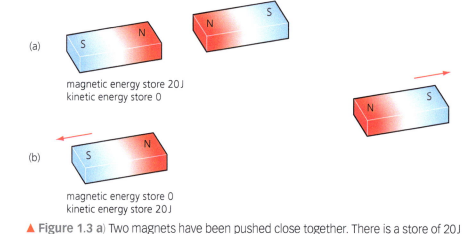

(a)

magnetic energy store 20 J
kinetic energy store 0

(b)

magnetic energy store 0
kinetic energy store 20 J

▲ **Figure 1.3 a)** Two magnets have been pushed close together. There is a store of 20 J of magnetic (potential) energy. **b)** The magnetic store of energy has reduced to zero, and each magnet has 10 J of kinetic energy, making a total store of 20 J.

The principle of conservation of energy

The **principle of conservation of energy** states that the amount of energy always remains the same. There are various stores of energy. In any process energy can be transferred from one store to another, but energy cannot be destroyed or created.

Transferring energy from one store to another

Light, sound and electricity are useful, but they are not stores of energy. They are ways of transferring energy from one store to a different energy store. You cannot go into a shop to buy a box of 'electrical energy', but you can buy a cell or battery. In a circuit, the chemical energy stored in a cell or battery causes electric charge to flow.

In a torch, the chemical energy stored in the battery causes an electric current (a flow of charge). The electric current causes the temperature of the bulb to increase so much that the bulb lights up. The light cannot be stored but it is useful. When the light strikes an object and is absorbed, the internal energy of the object increases.

If we drop a bunch of keys onto a table, the collision will make the air vibrate and we hear a sound. The sound wave transfers energy; it is not an energy store. The energy will transfer to the air and surrounding objects causing an increase in their store of internal energy.

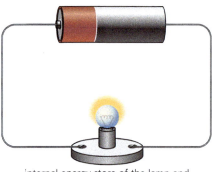

chemical energy
store reduces by 200 J

internal energy store of the lamp and
surroundings increases by 200 J

▲ **Figure 1.4** A cell passes a charge through a lamp. The charge flows for a few minutes. After this time the store of chemical energy in the cell has decreased by 200 J. The store of internal energy in the lamp and the surroundings has increased by 200 J.

Test yourself

1 Describe the energy stored in each of the following:
   a) a moving bicycle
   b) a compressed spring
   c) a bowl of breakfast cereal
   d) a rock lifted off the ground.
2 Explain what is wrong with this statement:
   'A car battery stores electrical energy for the lights, horn and starter motor.'
3 Describe how the stores of energy change from the beginning to the end of the following processes.

a) A catapult launches a marble.
b) A ball rolls along the ground and comes to rest.
c) A butane gas camping cooker heats up a pan of water.
d) A lump of soft putty falls to the ground.

4 Figure 1.5 shows a ball falling. Copy the diagram and fill in the values for the ball's kinetic energy and gravitational potential energy at each height.

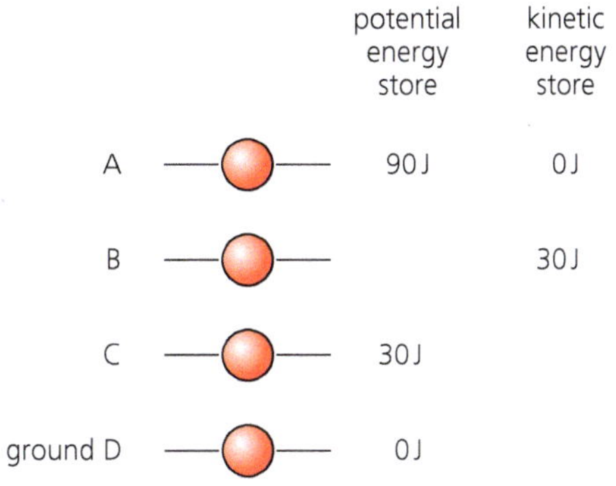

	potential energy store	kinetic energy store
A	90 J	0 J
B		30 J
C	30 J	
ground D	0 J	

▲ Figure 1.5

Show you can...

Complete this task to show you understand the different stores of energy.

Name four stores of energy, and describe three examples of how energy can be transferred from one store to another.

○ Calculating the energy

In this section you will learn how to calculate the amount of energy associated with a moving object, a stretched spring and an object raised above the ground. These calculations are useful to us. For example, we can show how the energy in a system is redistributed when a change happens to the system.

Kinetic energy

The kinetic energy stored by a moving object can be calculated using the equation:

$$E_k = \frac{1}{2} mv^2$$

$$\text{kinetic energy} = \frac{1}{2} \times \text{mass} \times (\text{speed})^2$$

Where energy is in joules, J
   mass is in kilograms, kg
   speed is in metres per second, m/s.

Elastic potential energy

The amount of elastic potential energy stored in a stretched spring can be calculated using the equation:

$$E_e = \frac{1}{2} ke^2$$

$$\text{elastic potential energy} = \frac{1}{2} \times \text{spring constant} \times (\text{extension})^2$$

Where energy is in joules, J
   spring constant is in newtons per metre, N/m
   extension is in metres, m.

The spring constant, k, is a measure of the spring's stiffness; k is equal to the force needed to stretch the spring one metre.

The extension of the spring is the increase in its length from its original unstretched length.

Gravitational potential energy

The amount of gravitational potential energy gained by an object raised above ground level can be calculated using the equation:

$$E_p = m\,g\,h$$

gravitational = mass × gravitational field strength × height
potential energy

Where energy is in joules, J

mass is in kilograms, kg

gravitational field strength is in newtons per kilogram, N/kg

height is in metres, m.

Example 1

A crate has a mass of 80 kg. A crane lifts the crate from a height of 3 m above ground to a height 18 m above the ground. Calculate the increase in gravitational potential energy of the crate.

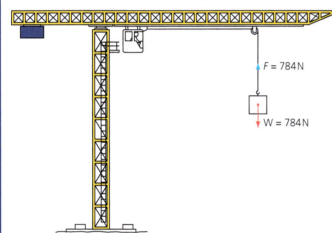

$F = 784\,\text{N}$

$W = 784\,\text{N}$

▲ **Figure 1.6** A crane does work to increase the potential energy of the crate.

Answer

Increase in height = 15 m

Increase in $E_p = mgh$

$\quad\quad = 80 \times 9.8 \times 15$

$\quad\quad = 11760\,\text{J or } 12\,\text{kJ (to 2 significant figures)}$

Example 2

A car has a kinetic energy store of 64 800 J. It is travelling at a speed of 12 m/s. Calculate the mass of the car.

Answer

$$E_k = \frac{1}{2}mv^2$$

$$64\,800 = \frac{1}{2} \times m \times (12)^2$$

$$m = \frac{2 \times 64800}{(12)^2}$$

$$m = 900\,\text{kg}$$

Example 3

A spring has a force constant of 60 N/m. The spring is extended by 5 cm. Calculate the elastic potential energy stored in the spring.

Answer

$$E_e = \frac{1}{2}ke^2$$

$$= \frac{1}{2} \times 60 \times (0.05)^2$$

$$= 0.075\,\text{J}$$

Remember to change the extension of 5 cm into metres.

Example 4

Calculate the change in kinetic energy stored when a car of mass 1200 kg slows down from 30 m/s to 20 m/s.

Answer

Change in kinetic energy $= \frac{1}{2}mv_1^2 - \frac{1}{2}mv_2^2$

$\quad\quad = \frac{1}{2} \times 1200 \times 30^2$

$\quad\quad - \frac{1}{2} \times 1200 \times 20^2$

$\quad\quad = 540\,000 - 240\,000$

$\quad\quad = 300\,000\,\text{J or } 300\,\text{kJ}$

Test yourself

5 Calculate the kinetic energy of a bullet of mass 0.015 kg, travelling at a speed of 240 m/s.
6 Calculate the increase in the gravitational potential energy store of a boy of mass 50 kg after he has climbed the Taipei 101 Tower, which is 440 m high.
7 A car has a mass of 1500 kg. It accelerates from a speed of 15 m/s to a speed of 20 m/s. Calculate the increase in kinetic energy of the car.
8 A car suspension spring has a spring constant of 2000 N/m. Calculate the elastic potential energy stored in the spring when it is compressed by 8 cm.
9 A meteor has a mass of 0.05 kg and it is travelling towards Earth at a speed of 30 km/s. Calculate the kinetic energy of the meteor.

○ Changes in energy

In this section you will learn how to use the energy equations to make predictions about changes to a system: which energy is transferred from one type of store to another.

Example 1

A ball of mass 100 g is thrown vertically upwards with a speed of 15 m/s. What is the maximum height the ball reaches?

Answer

As the ball rises, work is done by gravity to slow the ball down. Energy is transferred from the kinetic energy store of the ball to the gravitational potential energy store of the ball.

The principle of the conservation of energy tells us that the kinetic energy stored when the ball is at the bottom of its path is equal to the potential energy stored when the ball is at the top of its path, (if we assume that none of the energy is transferred to the surroundings).

$$\frac{1}{2}\,mv^2 = mgh$$
$$\frac{1}{2} \times 0.1 \times 15^2 = 0.1 \times 9.8 \times h$$
$$\text{So} \quad h = \frac{11.25}{0.98}$$
$$= 11.5\,\text{m}$$

Example 2

A stretched bow stores 64 J of elastic potential energy. The bow fires an arrow of mass 20 g. Calculate the speed of the arrow as it leaves the bow.

Answer

As the bow does work to speed up the arrow, energy is transferred from the bow's store of elastic potential energy to the arrow's store of kinetic energy.

$$\text{So} \frac{1}{2}\,mv^2 = 64$$
$$\frac{1}{2} \times 0.02 \times v^2 = 64$$
$$v^2 = \frac{64}{0.01}$$
$$v^2 = 6400$$
$$v = 80\,\text{m/s}$$

Example 3

A car of mass 1200 kg is parked on a 1 in 5 slope. The handbrake is released and the car rolls down the slope. When the car has travelled 20 m down the slope, how fast is it travelling?

Answer

A 1 in 5 slope means that the car goes up (or down) 1 m in height, when it travels 5 m along the slope. So, when the car has travelled 20 m along the slope it has gone down by a height of 4 m. The gravitational potential energy transferred can be calculated from this height.

$$mgh = \frac{1}{2}mv^2$$

gravitational field strength, $g = 9.8\,\text{N/kg}$

$$1200 \times 9.8 \times 4 = \frac{1}{2} \times 1200 \times v^2$$

$$\frac{1}{2}v^2 = 39.2$$

$$v^2 = 78.4$$

$$v = 8.9\,\text{m/s (to 2 significant figures)}$$

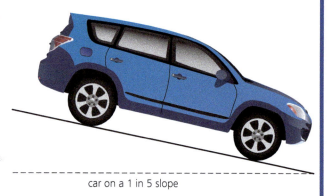

car on a 1 in 5 slope

▲ Figure 1.7

Test yourself

You can do these questions for practice, but you can also set up some experiments like this in the laboratory.

10 A spring with a spring constant of 200 N/m is fixed, so that it points vertically on to a table by using some Blu-Tack. The spring is compressed by 1 cm.
   a) Calculate the elastic potential energy stored in the spring.
      A polystyrene ball of mass 0.5 g is held on the spring and launched vertically upwards.
   b) i) How much gravitational potential energy does the ball have when it reaches its highest point?
      ii) Calculate the height the ball reaches when it is released. Assume the ball travels vertically upwards.

11 In Figure 1.9 a trolley is attached to a spring as shown. The trolley is pulled back to extend the spring by 15 cm. The spring has a spring constant of 80 N/cm.
   a) Calculate the elastic potential energy stored in the spring.
      The trolley is now released and it is accelerated by the spring.
   b) i) State the kinetic energy stored in the trolley just after the spring reaches its unstretched length.
      ii) Calculate the maximum speed of the trolley.

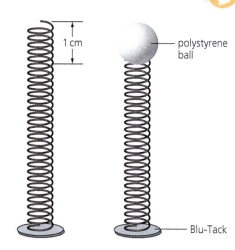

1 cm — polystyrene ball

Blu-Tack

▲ Figure 1.8

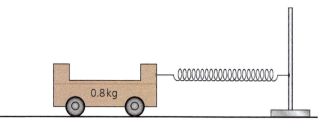

0.8 kg

▲ Figure 1.9

12 In Figure 1.10 a trolley is attached to a mass
    of 0.2 kg. The 0.2 kg mass is allowed to fall to
    the floor.
    a) Calculate the gravitational potential energy of
        the 0.2 kg when it is 0.9 m above the ground.
    b) i) State the kinetic energy of the 0.2 kg mass
            and the trolley together, just before the
            mass hits the ground.
        ii) Calculate the maximum speed of the trolley,
            just as the mass hits the floor.

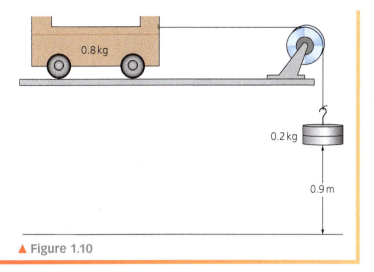

▲ Figure 1.10

Show you can...

State the principle of the conservation of energy. Describe two
demonstrations you have seen that help to explain this principle.

○ **Work**

KEY TERM

work = force × distance

In this section you are introduced to the definition of **work**, because
by doing work we can transfer energy from one store to another.

A force does work on an object when the force causes the object to move,
in the direction of the force. Work can be calculated using the equation:

$$W = Fs$$

work = force × distance moved in the direction of the force

Where work is in joules, J
    force is in newtons, N
    distance is in metres, m.

One joule of work is done when a force of 1 newton causes
a displacement of 1 metre.

1 joule = 1 newton – metre

When we do work, by applying a force to move an object, we change
the energy store of that object.

● When 200 J of work is done to lift a box upwards, the gravitational
  potential energy store of the box increases by 200 J.
● When 3000 J of work is done to accelerate a car, the kinetic energy
  store of the car increases by 3000 J.
● When 2 J of work is done to stretch a spring, the spring stores 2 J of
  elastic potential energy.

Electrical work is done by a battery when the battery makes a current
flow. (This is covered further in Chapter 2, p. 39)

←

Example

We can use the idea of work to help us calculate the braking distance of a car.

A car of mass 1500 kg is travelling at a speed of 20 m/s. The brakes apply a force of 5000 N to slow down and stop the car.

Calculate the braking distance of the car.

Answer

The decrease in the kinetic energy store of the car (transferred to the internal energy store of in the brakes) = work done against the braking force.

$$\frac{1}{2}mv^2 = Fs$$
$$\frac{1}{2} \times 1500 \times 20^2 = 5000 \times s$$
$$s = \frac{300\,000}{5000}$$
$$= 60\,\text{m}$$

◯ Power

Often when we want to do a job of work, we want to do it quickly. We say that a crane that lifts a crate more quickly than another crane lifting the same crate is more **powerful**.

Power is defined as the rate at which energy is transferred or the rate at which work is done. Power can be calculated by using these equations.

> **KEY TERM**
>
> $\text{power} = \dfrac{\text{energy transferred}}{\text{time}}$

▲ **Figure 1.11** When this train travels at 60 m/s, its engines run at a power of 2 MW.

$$P = \frac{E}{t}$$
$$\text{power} = \frac{\text{energy transferred}}{\text{time}}$$

$$P = \frac{W}{t}$$
$$\text{power} = \frac{\text{work done}}{\text{time}}$$

where power is in watts, W

energy transferred is in joules, J

time is in seconds, s

work done is in joules, J

An energy transfer of 1 joule per second is equal to a power of 1 watt.

Large powers are also measured in kilowatts, kW, and megawatts, MW.

$$1\,\text{kW} = 1000\,\text{W}\,(10^3\,\text{W}) \qquad 1\,\text{MW} = 1\,000\,000\,\text{W}\,(10^6\,\text{W})$$

Example

A weight lifter lifts a mass of 140 kg a height of 1.2 m in 0.6 s. Calculate the power developed by the weight lifter.

Answer

The potential energy transferred $= mgh$

$$E_p = 140 \times 9.8 \times 1.2 = 1646.4\,\text{J}$$

$$\text{power} = \frac{\text{energy transferred}}{\text{time}}$$

$$= \frac{1646.4}{0.6} = 2700\,\text{W or } 2.7\,\text{kW (to 2 significant figures)}$$

Practical

Measuring your own power

Work out your personal power by running up a flight of steps.
You need to know your mass, the time it takes you to run up the stairs and the vertical height of the stairs.

1 Record your time and calculate the increase in your gravitational potential energy store.
2 Now calculate your power.
3 Explain why you need the vertical height of the staircase and not the length along the staircase.

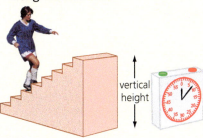

▲ **Figure 1.12** When you go up a flight of stairs, you are lifting your body weight and doing work against the force of gravity. The faster you run up the stairs, the greater your power.

TIP
Remember g = 9.8 N/kg.

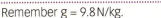

Test yourself

13 What is the unit of power?
14 What is the connection between the energy transferred and power?
15 A crane lifts a weight of 12 000 N through a height of 30 m in 90 s. Calculate the power output of the crane in kW.
16 Two students have an argument about who is more powerful. Peter says he is more powerful because he is bigger. Hannah says she is more powerful because she is quicker.
To settle the argument, they run up stairs of height 4.5 m. Use the information about their weights and times to settle the argument.

	Weight in N	Fastest time in s
Peter	760	3.80
Hannah	608	3.04

17 When an express train travels at a speed of 80 m/s, the resistive forces acting against it add up to 150 kN.
   a) Calculate the work done against the resistive forces in 1 s.
   b) Calculate the power output of the train, travelling at 80 m/s.

Show you can...

Design an experiment to measure the power of your arm as you lift a weight.

○ Energy changes in systems

You will find that some of this section is also covered in Chapter 3, page 74.

The amount of energy stored or released from a system, as its temperature changes, can be calculated using the equation:

$$\Delta E = mc\, \Delta\theta$$

change in thermal = mass × specific heat capacity × temperature change
energy

Where change in thermal energy is in joules, J

mass is in kilograms, kg

specific heat capacity is in joules per kilogram per degree Celsius, J/kg°C

temperature change is in degrees Celsius, °C.

The **specific heat capacity** of a substance is the amount of energy required to raise the temperature of one kilogram of the substance by one degree Celsius.

The specific heat capacity varies from substance to substance.

Required practical 1

An investigation to measure the specific heat capacity of a material

There are several different ways to obtain the data needed to calculate the specific heat capacity of a material. All of the methods involve the same idea; the decrease in one energy store (or work done) leads to an increase in the temperature of the material.

In this method energy is transferred from an electrical immersion heater to a metal block. The increase in the temperature of the metal block depends on the mass of the block and the specific heat capacity of the block.

Method

1 Measure the mass of the metal block (m) in kilograms.

2 Put the thermometer and immersion heater into the holes in the block.

3 Connect the immersion heater, joulemeter and power supply together as shown in Figure 1.13.

4 Measure the temperature of the metal block (θ_1) and then switch on the power supply.

5 Wait until the temperature of the block has gone up by about 10°C then switch off the power supply. Write down the reading on the joulemeter (E). This gives you the amount of energy transferred to the immersion heater.

6 Do not take the immersion heater out of the block. Keep looking at the temperature and write down the highest temperature shown by the thermometer (θ_2).

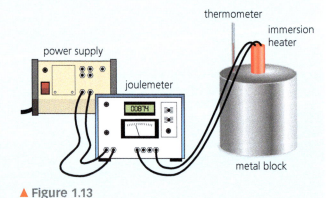

▲ Figure 1.13

If a joulemeter is not available set up the circuit shown in Figure 1.14. Use the following method to measure the energy transfer.

Switch the power supply on and again wait for the temperature of the block to increase by about 10 °C.

Watch the voltmeter and ammeter and write down the readings (*V* and *I*). The readings may change a little as the block gets warmer. Switch the power supply off and write down how many seconds the power supply was on for (*t*).

The energy transferred to the block can be calculated using the equation:

$$E = VIt$$

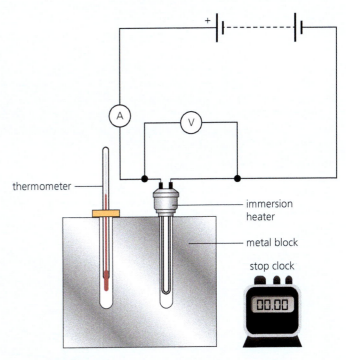

Analysing the results

1 Calculate the increase in temperature of the block $\Delta\theta = (\theta_2 - \theta_1)$.

2 Use the following equation to calculate the specific heat capacity (*c*) of the metal block:

$$c = \frac{E}{m\Delta\theta}$$

▲ **Figure 1.14**

3 It is likely that the value you calculate for the specific heat capacity of the metal will not be **accurate**.

Look up the true value. How close is the true value to your experimental value? Calculate the difference between the two values. Do you think your experimental value is accurate?

4 If other people in the class have used different types of metal, then compare the different specific heat capacity values with the temperature rise of the metal. If you do this, it is important that the blocks used by everyone have the same mass and that the same amount of energy is transferred to each immersion heater.

You should find that the higher the temperature rise, the smaller the specific heat capacity of the metal.

> **KEY TERM**
> A measurement or calculated value is considered **accurate** if it is close to the true value.

Taking it further

1 Repeat the investigation, but this time cover the side of the block in a thick layer of insulating material.

2 Calculate a second value for specific heat capacity. Is this second value any more accurate? If it is, suggest why.

Questions

1 Why is it better not to remove the immersion heater from the block as soon as the heater is switched off?

2 The values calculated for specific heat capacity from this investigation would usually be greater than the true value. Explain why.

Demonstration experiment

Your teacher might demonstrate how you can work out the temperature of a hot object.

(a)

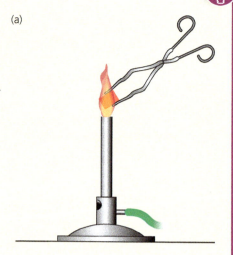

In Figure 1.15a) a small piece of steel is heated until it is red hot in a Bunsen flame. The steel is then quickly transferred to an insulated beaker, which contains 0.1 kg (100 ml) of water as shown in Figure 1.15b). *[Caution, the steel is red hot, and will burn the bench if dropped. The water will 'spit' as the steel is put into the beaker. Do not use a piece of steel of mass more than about 20 g.]*

Example data and calculation

Specific heat capacity of water	4200 J/kg °C
Specific heat capacity of steel	450 J/kg °C
Mass of steel	20 g
Temperature of water at the start of the experiment	19 °C
Temperature of water after the hot steel has been placed in it	34°C

(b)

The thermal energy transferred by the hot steel equals the thermal energy gained by the water.

The thermal energy gained by the water

$$= (mc\Delta\theta)_{\text{water}}$$

$$= 0.1 \times 4200 \times 15$$

$$= 6300 \text{ J}$$

Note: the temperature change is 34 °C – 19 °C = 15 °C

The thermal energy transferred by the hot steel is 6300 J.

▲ Figure 1.15

$$\text{So } 6300 = (mc\Delta\theta)_{\text{steel}}$$

$$6300 = 0.02 \times 450 \times \Delta\theta$$

$$\Delta\theta = \frac{6300}{0.02 \times 450}$$

$$\Delta\theta = 700 \text{ °C}$$

So the initial temperature of the steel was:

$$700 \text{ °C} + 34 \text{ °C} = 734 \text{ °C}$$

Note: the final temperature was 34 °C, which is added to 700 °C.

Give two reasons why the steel might have been even hotter than this.

Test yourself

18 State the units of specific heat capacity.

19 Air has a specific heat capacity of 1000 J/kg °C. Calculate the energy transferred by a heater when it warms the 80 kg of air in a room from 10 °C to 22 °C.

20 a) A night storage heater contains 60 kg of concrete, which has a specific heat capacity of 800 J/kg °C. How much energy must be supplied to the concrete to warm it from 15 °C to 45 °C?

   b) A 200 W heater is used to heat the concrete. How long does the heater take to supply the energy calculated in part (a)?

21 Milk has a specific heat capacity of 3800 J/kg °C. A cup of milk, of mass 0.3 kg, is placed into a microwave oven in a plastic insulating cup. When switched on the oven heats the milk at a rate of 700 W.

   a) Calculate how much energy has been transferred to the milk after 1 minute.

   b) When the milk was put in the oven its temperature was 6 °C. Calculate the temperature of the milk after 1 minute of heating.

Show you can...

Show you understand what is meant by specific heat capacity by completing this task.

Describe an experiment to measure the specific heat capacity of water. Explain what measurements you would take and the calculations you would do.

Conservation and dissipation of energy

KEY TERM

To **dissipate** means to scatter in all directions or to use wastefully. When energy has been dissipated, it means we cannot get it back. The energy has spread out and heats up the surroundings.

You have already learnt that energy is conserved: energy cannot be created or destroyed. However, when energy is transferred from a source, it is not all transferred usefully. Often when energy is transferred some of the energy is **dissipated** or 'wasted'.

When petrol is used to power a car, much of the energy is wasted. Only about a quarter of the chemical energy stored in the petrol is used to drive the car forwards (by doing work) and about three quarters of the chemical energy is wasted by heating up the engine.

◯ Reducing energy dissipation

Every day in our lives we use energy from fuels for transport or for heating our homes. In all cases the chemical energy stored in the fuels is eventually transferred to the thermal energy store of the surrounding area. This is a less useful way of storing energy; the energy is being 'wasted'. However, we try to ensure that as much energy is transferred usefully as possible. We try to minimise the amount of wasted energy.

Power stations

The purpose of a power station is to generate electricity, which can do useful work to light and heat our homes. Engineers design the generators in the power station to reduce the amount of waste energy in the power station. Generators are large machines which can dissipate energy by heating or by unwanted mechanical vibrations.

Car design

When we drive a car we want to make sure that as much of the chemical energy stored in the fuel as possible does useful work for us.

- Engineers design fuel efficient cars, which dissipate less energy.
- The car is made streamlined to reduce air resistance on the car.
- Moving parts of the car are lubricated with oil to reduce friction.

Keeping warm at home

When we heat our homes, the energy stored inside the house is dissipated through the roof, walls, windows or doors of the house to warm up the air outside. We want to make sure the energy escapes as slowly as possible.

Here are some ways in which we reduce unwanted energy dissipation at home.

Chimneys

Figure 1.16 shows a coal fire burning in a sitting room. Some of the energy from the burning coal is transferred to the air outside the house. This is wasted energy. By having the chimney inside the house, thermal energy can be transferred into the bedrooms upstairs. This is useful energy.

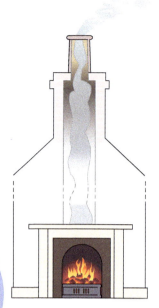

▲ Figure 1.16

Walls

The rate at which energy is transferred through the walls of a house depends on four factors.

- The temperature difference between inside and outside. (Our heating bills are larger in winter than in summer.)
- The area of the walls. (Large houses cost more to heat than small houses.)
- The thermal conductivity of the walls. Some materials conduct heat well, metals, for example. These materials have a high thermal conductivity. Brick and glass are not good thermal conductors. They have relatively low thermal conductivities, but energy still flows out of a warm house through the walls and windows. The higher the thermal conductivity of a material, the higher the rate of energy transfer by conduction across the material.
- The thickness of the walls (or windows) is important. The thicker the walls, the slower the rate of energy loss.

Modern houses are built with two layers of brick as shown in Figure 1.18. Then the house is insulated with cavity wall insulation, between the two layers of brick. The foam which insulates the walls is full of trapped air. The air is a good insulator; it has a much lower thermal conductivity than brick or glass.

▲ **Figure 1.17** This house was built in 1820. The walls are 70 cm thick. This keeps the house warm.

Loft insulation and carpets

The most efficient way to reduce energy loss from our house is to insulate the loft. A thick layer of loft insulation reduces energy loss through the roof. We also use insulating carpets to reduce energy loss through the floor.

The tiles on the kitchen floor feel cold when you walk on them in bare feet. These tiles are much better thermal conductors than carpets.

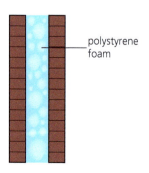

polystyrene foam

▲ Figure 1.18

Double glazing

A thin pane of glass in a window transfers energy out of the house. We use double glazing to reduce energy loss through the windows. A layer of gas trapped between two panes of glass provides good insulation (Figure 1.20).

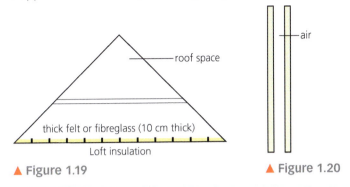

roof space

thick felt or fibreglass (10 cm thick)

Loft insulation

▲ Figure 1.19

air

▲ Figure 1.20

Investigating thermal insulation

Thermal insulation helps to keep things warm or cool for longer by reducing the unwanted energy transfers in a system.

This investigation is to compare how effective different materials are as thermal insulators.

Method
Risk assessment
- Hazard – using very hot water
- Risk – knocking the water over and scalding yourself
- Reducing the risk of harm – do not sit down while using the hot water
    – use hot water but not boiling water

1 Select different materials to test as insulators, for example aluminium foil, bubble wrap, thin expanded polystyrene, corrugated cardboard and cloth.

2 Wrap one of the materials around the outside of a metal or glass container (could be a 250 ml beaker). Do not wrap any material under the bottom of the container as this could make the container tip over.

3 Pour hot water into the container.

4 Put a cardboard lid onto the container and place a thermometer into the water through a hole in the lid. Make sure the thermometer is upright. (If it is at an angle it could tip the container over.) If necessary, hold the thermometer in a clamp stand.

5 Take the starting temperature of the water and start the stop clock.

6 Wait 10 minutes and then take the temperature of the water again.

7 Calculate the fall in temperature of the water.

8 Repeat this for each one of the materials being tested. To make it a **fair test**, each time you use a different material the volume of the water and the starting temperature of the water should be the same. These are both **control variables**.

Instead of a thermometer you could use a temperature sensor and data logger. If you do this it is likely that you will be able to measure the temperature to a greater **resolution** than if you use a thermometer.

Analysing the results
1 Draw a bar chart to show the temperature fall of the water for each of the materials tested.

2 Which material was the best insulator? Explain why it was better than the others.

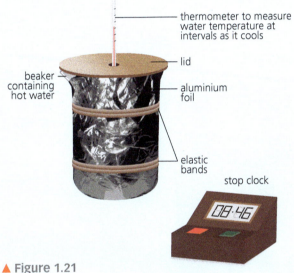

thermometer to measure water temperature at intervals as it cools

lid

beaker containing hot water

aluminium foil

elastic bands

stop clock

▲ Figure 1.21

KEY TERMS

In a **fair test** only the independent variable affects the dependent variable.

A **control variable** is what you keep the same.

The **resolution** of a measuring instrument is the smallest change that the instrument can detect.

In this investigation a bar chart is drawn because 'type of material' is a **categoric variable** and not a **continuous variable**.

Taking it further

Read the following hypothesis and prediction.

Hypothesis: The hotter a liquid, the faster the rate of energy transfer from the liquid.

Prediction: As a liquid cools down its temperature falls more slowly.

1 Plan and carry out an investigation to test the hypothesis and prediction.

2 How will you show that the data you collect supports or does not support the hypothesis?

Questions

1 Two ways of reducing the risk of harm are given in the risk assessment. What other precautions were used in the investigation to reduce the risk of scalding yourself with hot water?

2 Look at your results. Would a resolution of 1 °C on a thermometer be enough to be able to rank the materials you tested from best to worst insulator? Explain the reason for your answer.

KEY TERMS

A **categoric variable** has values that are given a name or a label.

A **continuous variable** has numerical values obtained by either measuring or by counting.

Test yourself

22 What is meant by the term 'energy dissipation'?

23 Explain how engineers can design cars to be more efficient.

24 Give two examples of how we reduce unwanted energy transfers. Try to think of two not mentioned earlier in the text.

25 List four ways that unwanted energy transfer is reduced in your home.

Show you can...

Show you understand the principle of conservation of energy by planning an experiment to demonstrate and prove this principle. State the apparatus you will use, the measurements you will take and the calculations you will do.

▲ **Figures 1.22 and 1.23** Both of these have done some useful work, but they have also wasted energy; they are inefficient.

○ Efficiency

Efficiency is a way of expressing the proportion of energy that is usefully transferred in a process as a number. The most efficient machines transfer the highest proportions of input energy to useful output energy.

To calculate efficiency we use the equation:

$$\text{efficiency} = \frac{\text{useful output energy transfer}}{\text{total input energy transfer}}$$

Efficiency may also be calculated using the equation:

$$\text{efficiency} = \frac{\text{useful power output}}{\text{total power input}}$$

Efficiency is a ratio of energies or powers. So efficiency has no unit. We write efficiencies as a decimal or a percentage.

Example 1

A steam engine uses coal as its source of energy. When the chemical energy store of the coal in the engine's furnace goes down by 150 kJ, the engine does 18 kJ of useful work against resistive forces.

Calculate the efficiency of the engine.

Answer

$$\text{efficiency} = \frac{\text{useful output energy transfer}}{\text{total input energy transfer}}$$

$$= \frac{18}{150}$$

$$= 0.12 \text{ or } 12\%$$

Note: here both energies were expressed in kJ. Make sure you use the same unit on the top and bottom of the fraction.

Example 2

A joulemeter records that 18.2 J of electrical work is done when an electric motor lifts a 0.3 kg load through a distance of 0.90 m. Calculate the efficiency of the motor.

Answer

$$\text{efficiency} = \frac{\text{useful output energy transfer}}{\text{total input energy transfer}}$$

$$= \frac{\text{gain in E}_p}{18.2} = \frac{mgh}{18.2}$$

$$= \frac{0.3 \times 9.8 \times 0.90}{18.2}$$

$$= 0.15 \text{ or } 15\% \text{ (2 significant figures)}$$

(H) Increasing the efficiency of an intended energy transfer

Whenever we do a job of work, we want to ensure that as much energy as possible is transferred usefully, and that little energy is dissipated wastefully.

When we move an object we transfer energy by doing work, and work is calculated using the equation:

$$W = Fs$$

where F is the force applied and s is the distance moved in the direction of the force. We do less work if the forces of friction or air resistance that act against us are small. When frictional forces act, energy is transferred to the thermal energy store of the surroundings. This is wasteful.

We reduce friction by:

- using wheels
- applying lubrication.

We reduce air resistance by:

● travelling slowly
● streamlining.

When we streamline a car (for example), we are shaping its surface so that air flows past the car and offers as little resistance to the motion of the car as possible.

Increasing efficiency using machines

We can also increase the efficiency of a job by using a machine. Figure 1.24 shows two men lifting a load of bricks on a building site. The man on the ground pulls with a force of 250 N. Because there are four ropes in their system of pulleys, the ropes apply a force of 4 × 250 N (1 000 N), which is enough to lift the bricks and the pulley, and to overcome friction.

Without the machine the men would have to carry the bricks up the ladder in smaller loads. This means that they would have to work to carry the bricks and to lift their own weight. The machine allows them to apply a smaller force to lift the bricks, but they have to pull the rope 4 times as far. Using the machine is far more efficient than climbing up and down the ladder. It saves time and is much safer.

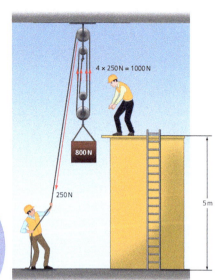

▲ Figure 1.24

Test yourself

26 Define efficiency.

27 A student compares two machines, A and B. Both machines transfer the same input energy. The student discovers that machine A wastes less energy than machine B. Which machine is more efficient?

28 Phil is in the gym doing pull-ups. Each time he does a pull-up his store of chemical energy decreases by 1500 J. Phil's mass is 72 kg and he lifts himself up 0.5 m in one pull-up.

   a) Calculate the gravitational potential energy stored after one pull-up.

   b) Calculate the efficiency of Phil's body during this exercise.

29 A car's engine is supplied with one kilogram of fuel, which stores 45 MJ of chemical energy. The efficiency of the car is 36%.

   Calculate the amount of energy available for useful work against resistive forces.

30 Why does a streamlined car use fuel more efficiently than another similar car which experiences larger air resistance forces?

31 a) Why does the machine in Figure 1.25 allow the men to work more efficiently?

   b) Suggest another machine that increases the efficiency of a job. How does the machine ensure that more energy is usefully transferred?

▲ Figure 1.25

Show you can...

Show that you understand the meaning of the word *efficiency* by designing an experiment to determine the efficiency of an electric motor. Explain what measurements you would take and what calculations you would do.

National and global energy resources

Every day we depend on various energy resources to make our lives comfortable. A hundred and fifty years ago our ancestors walked to school, lived in cold houses and did not have electricity in their homes. Now we travel by car, train or bus, we live in warm homes, and all of us use electricity to run the many appliances we have at home.

◯ Fossil fuels

Much of our energy in the UK comes from the fossil fuels, coal, oil and gas. Like all fuels, they store energy. However, to release the energy, the fossil fuels must be burned. Once the fuels are burnt, they are gone forever, because fossil fuels have taken millions of years to be formed. Fossil fuels are described as **non-renewable energy resources**, because there is a finite supply of them.

By contrast **renewable energy resources** will never run out. We can obtain renewable energy from the Sun, tides, waves and rivers, from the wind and from the thermal energy of the Earth itself.

Finite resources

Figure 1.26 shows that we only have relatively small supplies of fossil fuels left. Unless we find substantial new resources, the world's supply of oil and gas will run out by about 2070 and coal will run out by about 2130. Our descendants will have to use different energy resources.

◯ Using fuels

In the UK the three main fossil fuels provide most of the energy for our needs.

- **Transport.** Fuels such as petrol, diesel and kerosene are produced from oil. These fuels drive our cars, trains and planes. Electricity is also used to run our trains, and cars are being developed to run from electricity supplies. In 50 years' time, we may not be able to fly on holiday.
- **Heating.** Most of our home heating is provided by gas and electricity. Gas pipes run into our houses to provide energy to our boilers. Some homes are warmed by oil-fired boilers or by burning solid fuels such as coal and wood.
- **Electricity.** In the UK, electricity is generated using different energy resources, but most of our electricity is generated by burning fossil fuels. The following table shows the percentage of electricity generated using different energy resources. Gas and coal provide a convenient and relatively cheap way to generate our electricity. However, the burning of gas and coal produces carbon dioxide, which most scientists think is responsible for global warming. In November 2015, the UK government announced that coal fuelled power stations will be phased out by 2025. More electricity will be generated by gas, which produces less carbon dioxide than coal.

KEY TERMS ⭐

Non-renewable energy resources
These will run out, because there are finite reserves, which cannot be replenished.

Renewable energy resources
These will never run out. They are (or can be) replenished as they are used.

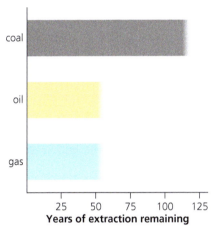

▲ **Figure 1.26** The world's supply of fossil fuels will not last forever. The information here shows current estimates based on known reserves; many more may be discovered.

Percentage of UK electricity generated by different energy resources.
(Source: Department of Energy and Climate Change, September 2015.) We can expect a greater amount of electricity to be generated by renewable sources in future.

Energy resource used to generate electricity	Percentage of UK electricity generated from each energy resource 2015
Gas	30
Coal	20
Nuclear	22
Biomass	9
Wind	11
Hydroelectric	2
Oil	2
Solar	4

▲ **Figure 1.27** Trees killed by acid rain.

KEY TERM ⭐

1 GW, 1 gigawatt = 10^9 W

Fossil fuels and acid rain

One of the products of burning coal is sulfur dioxide. When sulfur dioxide combines with water, acid rain is produced. Acid rain damages buildings and kills plants.

Sulfur dioxide can be removed from the waste gases of burning coal, but this is expensive.

Power stations

The average consumption of electrical power in the UK is about 36 GW and our peak consumption (in the evening) reaches 57 GW. [**1 GW** is 1 gigawatt, which is 1000 million watts or 10^9 watts.]

Figure 1.28 shows the layout in a coal-fired power station. You are not expected to remember this for your GCSE but it is helpful to understand the principle behind generating electricity: an energy resource such as gas or water moves past a turbine; the turbine then drives the generator.

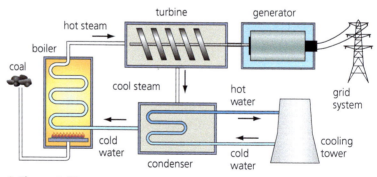

▲ **Figure 1.28**

○ **Other energy resources**

Due to concerns over global warming, which is caused (in part) by the burning of fossil fuels, governments must find alternative energy resources. In 2008, the UK government committed itself to reducing carbon dioxide emissions by 80% by 2050. In 1990 only about 4% of the UK's electricity generation used renewable resources. Now that figure has increased to 26%.

Nuclear power

Nuclear power generates about 22% of the electricity in the UK. There are currently plans for a new nuclear power station at Moorside in West Cumbria. The Moorside power station will begin generating electricity in 2024. The peak generating capacity of the power station will be 3.4 GW.

The nuclear fuels used are mainly uranium and plutonium. These are also non-renewable energy resources. However, nuclear fuel contains a huge amount of nuclear energy, and it is estimated that there is enough uranium to last thousands of years.

Nuclear power has the advantage of producing no pollutant gases. However, we need to be very careful how we store nuclear waste.

Biomass

Waste products can provide fuel for some small electrical generators. Much of the waste is wood, so this is a renewable energy resource. Many of the biomass generators are also used to heat factories or houses directly.

Biofuels emit carbon dioxide when they are burnt. However, the plants that become biofuels used carbon dioxide as they grew. So, overall biofuels do not add to the amount of carbon dioxide in the atmosphere. They are '**carbon neutral**'.

Using tides

Every day tides rise and fall. Massive amounts of water move in and out of river estuaries. It is estimated that the energy of the tides could generate about 20% of Britain's electricity.

At present there is little electricity generated by tidal energy. Expensive barriers must be constructed. The largest tidal barriers in the world generate about 250 MW of electricity.

A barrage is like a dam built across a river estuary (Figure 1.29). The barrage has underwater gates that open as the tide comes in and then close to keep the water behind the barrage (Figure 1.30). When the tide goes out, a second set of gates is opened. Water flows out of these gates and drives turbines that are connected to generators as it does so.

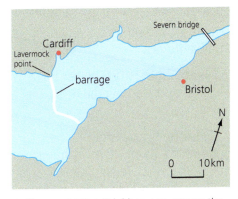

▲ **Figure 1.29** A tidal barrage across the River Severn could generate about 6% of Britain's electricity. The peak power would be about 7 GW.

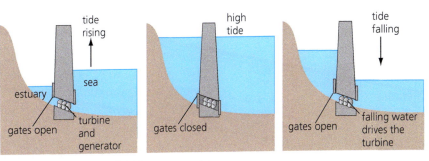

▲ **Figure 1.30** How a tidal barrage works.

Hydroelectric power

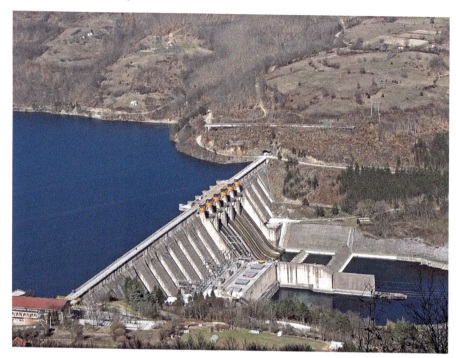

▲ **Figure 1.31** Hydroelectric power stations can be huge and generate vast amounts of power.

Hydroelectric power stations generate about 2% of Britain's electricity, but they generate about 10% of the world's electricity.

Many hydroelectric power schemes have had environmental impacts, as a new lake is formed.

- Forests have been cut down.
- Farmland is lost.
- Wildlife habitats have been destroyed.
- Many people have had to move homes.

Wind power

When the wind blows with sufficient force, the blades of a wind turbine rotate. The blades turn a generator which produces electricity. Wind power has the advantage of being clean. There are no waste products. However, wind power also has some disadvantages.

- Wind power is unreliable. If the wind is too light, little power is produced. If the wind is too strong, generators can overheat, so the blades have to be stopped from moving.

At present 11% of Britain's electricity comes from wind power. The government's target is to increase this figure to about 20%. Although wind power is unreliable, when the wind does blow, we can turn off gas generators. So wind power will save fossil fuels and reduce greenhouse gas emissions.

a) turbine blade, transmission shaft, tower, generator

▲ **Figures 1.32a)** and **b)** In 2010 the UK government approved the building of 6000 new wind turbines to be erected off the coast of Britain. This will give a total of 12000 turbines with a maximum power capacity of 32GW.

Solar power

Solar cells use energy directly from the Sun to generate electricity. Solar cells generate electricity on a small scale.

In some countries solar electricity generation is unreliable, due to the weather, but it can still make a useful contribution to a country's overall power supply.

Geothermal energy

A small number of countries are able to make good use of geothermal energy to generate electricity. Iceland generates 30% of its electricity by taking advantage of the volcanic activity on the island. In some cases hot water is used directly to warm houses. Hot water or steam is also used to generate electricity. This form of electricity generation has the advantage of being clean and renewable.

▲ **Figure 1.33** This traffic slowing sign is powered by solar cells.

▲ **Figure 1.34** Solar panels on a house.

▲ **Figure 1.35** Generating energy from geothermal sources.

Test yourself

32 a) What is a non-renewable energy resource? Give one example.
  b) What is a renewable energy resource? Give one example.
33 Name a common fuel used in a nuclear power station. Is this fuel renewable or non-renewable?
34 In Britain 20% of our electricity is generated in coal-fired power stations.
  a) What are the advantages of using coal to generate electricity?
  b) State two environmental problems caused by using coal to generate electricity.
35 a) What environmental problems are caused by building a hydroelectric power station?
  b) Give one advantage of hydroelectric power.
36 Why are tides a more reliable way of generating electricity than wind power?
37 Britain plans to have 12 000 wind turbines spread from the south to the north of the country. Why does spreading out the wind farms increase the reliability of wind power?

38 A wind turbine is designed to produce a maximum power of 4 MW. However, due to variations of wind speed, the generator only produces this power for 10% of the time.
How many such wind turbines are required to replace a coal-fired power station which generates 2000 MW of power all the time?

39 Figure 1.36 shows the layout of a pumped storage power station. Water from the high level lake generates electricity by flowing through the turbines which are connected to generators. These are placed above the low level lake. When there is a low demand for electricity, the generators are driven in reverse to pump water back into the high level lake. This ensures there is enough water to generate electricity next time the demand is high.

a) Why is this sort of power station useful to electricity companies?

b) Does this power station produce any pollution or greenhouse gases
i) when generating electricity, ii) when pumping water back up the hill?

c) Use the information in Figure 1.36 to calculate the gravitational potential energy transferred per second when the generators are working.

d) The generators are 80% efficient. Calculate the power output of the power station in MW.

Figure 1.37 shows the typical power use in the UK on a spring day.

e) At what time of the day does the pumped storage station i) generate electricity, ii) pump water back up the hill? Give reasons for your answer.

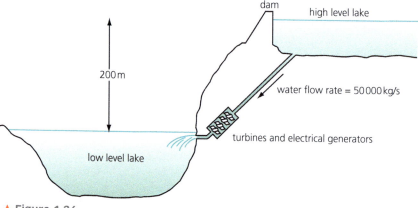

▲ Figure 1.36

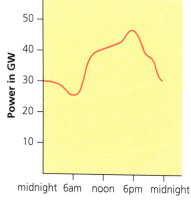

▲ Figure 1.37 A typical day's use of electricity on a spring day in the UK.

Show you can...

Show you understand the issues that affect the generation of electricity by writing a plan for electricity generation in the UK for 2040. Which types of generation are you going to use, and where will you site your power stations? Your brief report should give proper consideration to ethical issues.

Energy: a summary

In this book we provide clear definitions using 'key terms' where possible. No such easy definition exists for energy. Instead we have included this brief summary to pull the ideas about energy together.

Energy is an idea that cannot be described by a single process, nor is energy something we can hold or measure directly. However, we pay an enormous amount of attention to energy because it is conserved.

There are many different stores of energy. In any process, energy can be transferred from one store to another store, but energy is never destroyed or created.

Energy stores

You have met the following stores of energy:

- kinetic
- chemical
- internal (or thermal)
- gravitational potential
- magnetic
- electrostatic
- elastic potential
- nuclear.

Energy transfers

There are various ways that energy can be transferred:

- by mechanical work
- by electrical work
- by heating
- by radiation.

The word 'radiation' includes light and all electromagnetic waves. Radiation also includes 'mechanical radiation' such as sound and shock waves. Sometimes people refer to light, sound and electrical energy. However, light, sound and electricity are not energy stores, they transfer energy from one store to another. Here are some examples.

A hot cup of tea has a store of thermal energy. As the tea cools, it heats the surroundings. Energy is transferred to the thermal store of the surroundings and the temperature of the surroundings goes up (but is too small to measure).

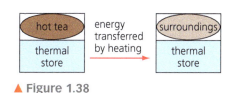

▲ Figure 1.38

A battery stores chemical energy. Energy is transferred by electrical work to the lamp. The lamp does not store energy. The lamp transfers energy to the thermal store of the surroundings by heating (conduction and convection) and radiation. Some of this radiation is useful to us as it is transferred; this is visible light.

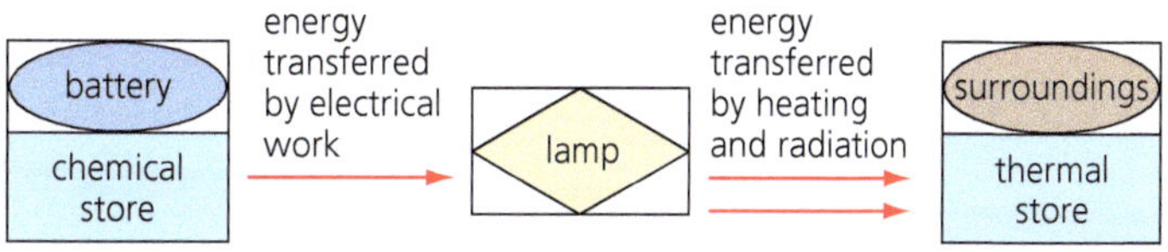

▲ Figure 1.39

A motor is used to lift a mass. A battery stores chemical energy. Energy is transferred by electrical work from a battery to a motor. The motor does not store energy, but it has a temporary store of kinetic energy when it is turning. The motor does mechanical work and transfers energy to the gravitational potential store of the raised mass. The motor also transfers energy to the thermal store of the surroundings by heating and by making a noise.

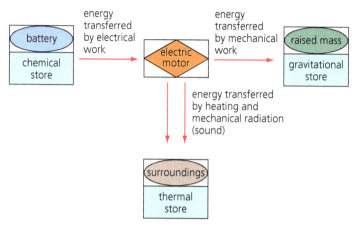

▲ Figure 1.40

And finally...

Energy can be transferred from one energy store to a different store, and there is as much energy stored at the end of the process as there was at the beginning. The reason energy is useful to us is that it allows us to do various jobs of work and to keep warm.

Chapter review questions

1 Describe the energy store of each of the following:

   a) an electrical battery

   b) a moving car

   c) a stretched rubber band

   d) a lake of water behind a hydroelectric dam.

2 In the following processes energy is transferred from one energy store to another store or stores. State the stores of energy at the beginning and end of each process.

   a) A battery lights a lamp.

   b) A bowl of hot soup cools.

   c) A battery is connected to a motor which is lifting a load.

   d) A firework rocket has been launched into the air and is travelling upwards.

3 Your feet feel warmer on a carpet than they do a tiled kitchen floor. Give one reason why.

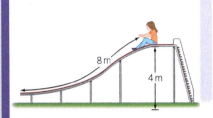

▲ Figure 1.41

4 Calculate the energy stored in each of these examples.

   a) A car of mass 1400 kg travels at a speed of 25 m/s.

   b) A suspension spring for a truck with a spring constant of 40 000 N/m is compressed by 5 cm.

   c) A suitcase of mass 18 kg is placed in a luggage rack 2.5 m above the floor of a train.

5 Figure 1.41 shows a girl on a slide. Her mass is 45 kg.

   a) Calculate the girl's speed at the bottom of the slide. Ignore the effects of friction.

   b) Explain why the girl's speed is likely to be less than the answer calculated in part (a).

6 A gymnast of mass 55 kg lands from a height of 5 m onto a trampoline (see Figure 1.42). Calculate how far the trampoline stretches before the gymnast comes to rest.
The trampoline has a spring constant of 35 000 N/m.

▲ Figure 1.42

7 Some lead shot with a mass of 50 g is placed into a cardboard tube as shown in Figure 1.43. The ends of the tube are sealed with rubber bungs to keep the lead shot in place. The tube is rotated so that the lead shot falls and hits the bung at the bottom.

   a) Why does the temperature of the lead shot increase after it has fallen and hit the lower bung?

   b) A student rotates the tube 50 times. Calculate the total decrease in the gravitational potential energy store of the lead shot in this process.

   c) The specific heat capacity of lead is 160 J/kg °C. Calculate the temperature rise of the lead shot after the student has rotated the tube 50 times.

   d) In practice, the temperature rise of the lead shot is likely to be less than your answer in part (c). Give a reason why.

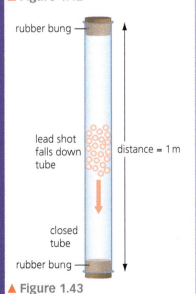

▲ Figure 1.43

8 A girl kicks a football with a force of 300 N. The girl's foot is in contact with the ball for a distance of 0.2 m. The ball has been a mass of 450 g. Calculate the speed of the ball just after it has been kicked.

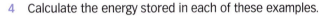

9 An electricity supply provides electrical power to a motor at a rate of 800W. The motor lifts a crate of mass 80 kg through a height of 3 m in 12 seconds. Calculate the efficiency of the motor.

10 A student wants to calculate the specific heat capacity of a liquid. First he determines that the liquid has a density of 900 kg/m^3.

a) The student places a volume of 200 cm^3 of liquid into an insulated beaker. Calculate the mass of the liquid in kg. [1 cm^3 = 10^{-6} m^3]

The student measures the temperature of the liquid. It is 22 °C. The student then heats the liquid with a heater that has a power rating of 24W. He heats the liquid for 10 minutes.

b) Calculate the energy transferred to the liquid in 10 minutes.

c) After 10 minutes the student finds the liquid has risen to a temperature of 72 °C. Calculate the specific heat capacity of the liquid.

11 A scientist observes a grasshopper as it jumps. The grasshopper takes 25 milliseconds to take off, and reaches a speed of 3.0 m/s. The grasshopper has a mass of 1.5 g.

a) Calculate the kinetic energy of the grasshopper after its jump.

b) Calculate the mechanical power developed by the grasshopper's legs.

Practice questions

1 Which of the following is the correct unit for power?

newtons   joules   watts   [1 mark]

2 Which of the following is the correct unit for specific heat capacity?

J/kg°C   J kg/°C   J kg°C   [1 mark]

3 The energy input to a machine is 2000 J. The machine transfers 600 J of useful energy. Calculate the efficiency of the machine.   [2 marks]

4 The British government has planned to build up to 12 000 wind generators. Give one advantage of wind power and one disadvantage.   [2 marks]

5 Many of the world's electricity power stations burn fossil fuels.

a) Burning fossil fuels produces carbon dioxide.

What effect can an increase in carbon dioxide levels have on the Earth's atmosphere?   [1 mark]

b) Figure 1.44 shows how much carbon dioxide is produced for each unit of electricity generated in coal-, gas- and oil-burning power stations.

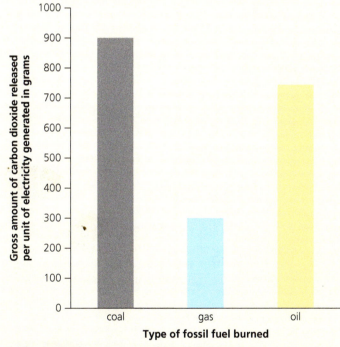

▲ Figure 1.44

i) Which type of fossil fuel produces the most carbon dioxide for each unit of electricity generated?   [1 mark]

ii) Why is a bar chart drawn to show the data and not a line graph?   [1 mark]

c) Biofuels are renewable energy resources that can be used to generate electricity.

i) Name one other type of renewable energy resource.   [1 mark]

ii) Most biofuels are derived from plants. When biofuels are burned carbon dioxide is produced. Why does burning a biofuel have less effect on the atmosphere than burning coal?   [1 mark]

6 An advertisement for solid fuel firelighters claims:

▲ Figure 1.45

a) To test this claim a student plans the following investigation.

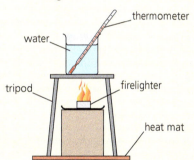

▲ Figure 1.46

- Place 1 g of H&S firelighter on a tin lid.
- Put 80 cm³ of water into a beaker.
- Measure the temperature of the water.
- Set fire to the firelighter and then use it to heat the water.
- When all of the firelighter has burned, measure the new water temperature.
- Repeat with two different brands of firelighter.

i) What type of variable is the brand of firelighter?   [1 mark]

ii) Name two control variables in this investigation.   [2 marks]

iii) Give one experimental hazard in this investigation.   [1 mark]

iv) Suggest one change that the student could have made to improve the resolution of the temperature readings.   [1 mark]

b) To compare the data the student drew the bar chart shown in Figure 1.47.

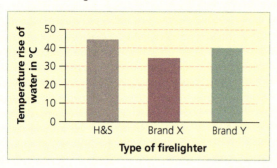

▲ Figure 1.47

i) Was the data collected by the student sufficient to confirm the claim made by the maker of H&S firelighters? Give a reason for your answer. [2 marks]

ii) Give **two** reasons why the decrease in the chemical energy store of the firelighter is greater than the increase in the thermal energy store of the water. [2 marks]

7 The roof of the Tokyo Skytree is 495 m high and can be climbed using its steps. A tourist decided to climb the tower. He took 35 minutes to do it and he had a mass of 60 kg.

a) Calculate the increase in the gravitational potential energy store of the tourist in climbing the steps. [3 marks]

b) Calculate his average power output during the climb. [3 marks]

c) The energy to make the climb is transferred from the chemical energy store of the tourist. Eating one slice of bread provides 400 kJ of chemical energy. Calculate the number of slices of bread he should eat for breakfast to provide the energy for the climb.

(Assume his body is 20% efficient at transferring energy from his chemical energy store to gravitational potential energy.) [3 marks]

d) Where is most of the energy from the tourist's food transferred to? [1 mark]

8 In Figure 1.48 a conveyor belt is used to lift bags of cement on a building site.

▲ Figure 1.48

a) A 40 kg bag of cement is lifted from the ground to the top of the building. Calculate its gain in gravitational potential energy. [3 marks]

b) The machine lifts five bags per minute to the top of the building. Calculate the useful energy delivered by the machine each second. [2 marks]

c) The machine is 35% efficient. Calculate the input power to the machine while it is lifting the bags. [3 marks]

9 A car and its passengers have a combined mass of 1500 kg. The car is travelling at a speed of 15 m/s. It then increases speed to 25 m/s.

Calculate the increase in kinetic energy of the car. [3 marks]

10 Figure 1.49 shows a Pirate Boat theme park ride which swings from A to B to C and back.

a) As the boat swings from A to B a child increases her kinetic energy store by 10 830 J. The child has a mass of 60 kg and sits in the centre of the boat. Calculate the speed of the child as the boat passes through B. [3 marks]

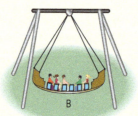

▲ Figure 1.49

b) Sketch a graph to show how the child's gravitational potential energy changes as the boat swings from A to B to C. [3 marks]

c) Calculate the change in height of the ride.

(Assume the decrease in the gravitational potential energy store as the child falls is transferred to the kinetic energy store of the child.) [3 marks]

11 An electric winch is used to pull a truck up a slope, as shown in Figure 1.50.

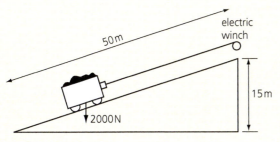

▲ **Figure 1.50**

a) How much work is done in lifting the truck 15 m? [3 marks]

The winch uses a 6 kW electrical supply, and pulls the truck up the slope at a rate of 5 m/s.

b) How long does it take to pull the truck up the slope? [2 marks]

c) How much work is done by the winch? [2 marks]

d) Calculate the efficiency of the winch. [2 marks]

12 A barrage could be built across the estuary of the River Severn. This would make a lake with a surface area of about 200 km² (200 million m²).

Figure 1.51 shows that the sea level could change by 9 m between low and high tide; but the level in the lake would only change by 5 m.

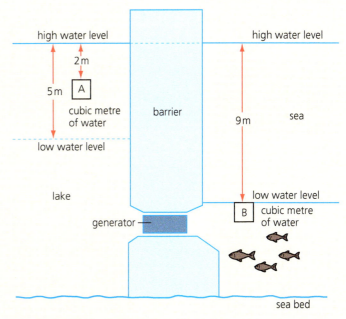

▲ **Figure 1.51**

a) Calculate the gravitational potential energy transferred when a mass of 1 kg falls from A to B. [3 marks]

b) Calculate the number of cubic metres of water that flow out of the lake between high and low tides. [2 marks]

c) Calculate the mass of water that flows out between high and low tides. A cubic metre of water has a mass of 1000 kg. [2 marks]

d) Use your answers to parts (a) and (c) to calculate how much energy can be transferred from the tide. Assume that position A is the average position of a cubic metre of water between high and low tide. [3 marks]

e) The time between high and low tide is approximately 6 hours. Use this figure to calculate the average power available from the dam. Give your answer in megawatts. [3 marks]

f) What are the advantages and disadvantages of the Severn barrage as a possible source of power? [4 marks]

Working scientifically

Uncertainty, errors and precision

When an electric motor is used to lift a weight, the power supply does electrical work to make the motor turn.

The motor transfers some energy usefully. This increases the gravitational potential energy stored by the weight. The rest of the energy is dissipated to the surroundings.

Susan decided to find out if the efficiency of an electric motor depends on the size of the weight being lifted. To do this she set up the apparatus shown in Figure 1.52.

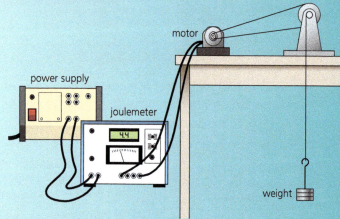

▲ Figure 1.52

Susan started with a 2 N weight. She switched on the power supply and increased the potential difference (p.d.) until the motor just lifted the weight from the floor to the bench top; a distance of 0.8 m. The joulemeter recorded the energy transfer to the motor. Susan repeated this step twice more. Her results are recorded in the table.

Trial	Joulemeter reading
1	14.8
2	15.3
3	14.9

1 Show that the mean (average) joulemeter reading was 15.0.

The three readings taken from the joulemeter are all close to the mean. The values for the energy transferred are precise.

When the same quantity is measured several times, the bigger the spread of the measurements about the mean, the less precise the measurements are. The precision of a set of measurements depends on the extent of the random errors.

The uncertainty in these energy values is ±0.3. The uncertainty is worked out by calculating the difference between the mean value and the value furthest away from the mean (in this case 15.3 – 15.0).

An uncertainty in a set of data can be caused by random errors or a systematic error. In this investigation, judging when the weight reaches the bench top and then switching off the power supply is a random error. If whenever the joulemeter is reset it does not go back to zero, this is a systematic error. This type of systematic error is also called a zero error.

KEY TERMS

A **precise** set of measurements of the same quantity will closely agree with each other.

A **random error** is unpredictable. Repeating the measurement and then working out a mean will reduce the effect of a random error. In a graph, data that is scattered about the line of best fit shows a random error.

For a set of measurements, the difference between the maximum value and the mean or the minimum value and the mean gives a measure of the **uncertainty** in the measurements.

A **systematic error** is a consistent error usually caused by the measuring instruments. All of the data is higher or lower than the true value. Data with a systematic error will give a graph line that is higher or lower than it should be.

A **zero error** is when a measuring instrument gives a reading when the true value is zero.

Susan repeated the procedure with a range of different weights. For each weight, she obtained three energy values and recorded the mean. These values are shown in the table.

Mean energy input to the motor in J	Weight lifted in N	Percentage efficiency of the motor
15.0	2	10.7
16.9	3	14.2
21.2	4	
25.5	5	15.7
33.5	6	14.3
46.0	7	12.2

2 In Susan's investigation:

a) What range of weights was used?

b) What was the independent variable?

c) Which variable was controlled during the investigation?

3 How could Susan tell that the joulemeter she used did not have a zero error?

4 Copy and complete the table above by calculating the missing efficiency value.

5 Draw a graph of percentage efficiency against weight lifted.

6 What can you conclude from this investigation?

2 Electricity

At the beginning of the twentieth century, very few people had electricity supplied to their homes. Now, a century later, the supply of electricity to our homes, offices and streets is an essential part of life. However, electricity comes at a price and we must be careful how much we use it. We must plan ahead to make sure we are able to provide electricity far into the future.

Specification coverage

This chapter covers specification points: 4.2.1 Current, potential difference and resistance, 4.2.2 Series and parallel circuits, 4.2.3 Domestic uses and safety and 4.2.4 Energy transfers.

Prior knowledge

Previously you could have learned

› When two objects rub against each other, electrons can transfer from one object to another. When electrons transfer to an object it becomes negatively charged. When electrons leave an object it becomes positively charged.

› Two like charges repel each other.

› Two unlike charges attract each other.

› An electrical current is a flow of charge.

› In metals, current is a flow of electrons.

› Some materials are good conductors of electricity.

› Some materials do not conduct electricity. These are called insulators.

› A cell or battery has a store of chemical energy. The energy stored decreases when a charge flows.

› When the same current passes through a number of components they are said to be in **series**.

› In a **parallel** circuit, the current divides into different branches.

Test yourself on prior knowledge

1 Are the lights in your home in series or in parallel? How can you tell?

2 a) Name three good conductors of electricity.

   b) Name three good insulators of electricity.

3 A piece of plastic is rubbed with a cloth. The plastic becomes negatively charged.

   a) Explain how the plastic has become charged.

   b) What type of charge is now on the cloth?

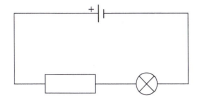

▲ **Figure 2.1** The resistor and lamp are in series.

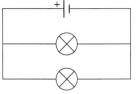

▲ **Figure 2.2** The two lamps are in parallel.

Current, potential difference and resistance

○ Circuit symbols

Figure 2.3 shows the circuit symbols for the electrical components that you will meet in this section. A brief explanation of their function is given here and you will learn more about them later on.

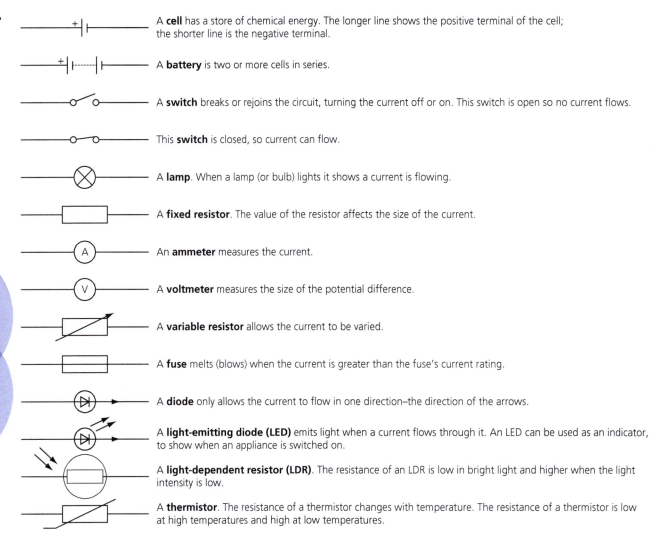

A **cell** has a store of chemical energy. The longer line shows the positive terminal of the cell; the shorter line is the negative terminal.

A **battery** is two or more cells in series.

A **switch** breaks or rejoins the circuit, turning the current off or on. This switch is open so no current flows.

This **switch** is closed, so current can flow.

A **lamp**. When a lamp (or bulb) lights it shows a current is flowing.

A **fixed resistor**. The value of the resistor affects the size of the current.

An **ammeter** measures the current.

A **voltmeter** measures the size of the potential difference.

A **variable resistor** allows the current to be varied.

A **fuse** melts (blows) when the current is greater than the fuse's current rating.

A **diode** only allows the current to flow in one direction–the direction of the arrows.

A **light-emitting diode (LED)** emits light when a current flows through it. An LED can be used as an indicator, to show when an appliance is switched on.

A **light-dependent resistor (LDR)**. The resistance of an LDR is low in bright light and higher when the light intensity is low.

A **thermistor**. The resistance of a thermistor changes with temperature. The resistance of a thermistor is low at high temperatures and high at low temperatures.

▲ **Figure 2.3** Circuit symbols.

Show you can...

Complete this task to show that you understand how to draw and design electrical circuits. Draw a circuit diagram to show how two lamps can be connected to a battery, with components that allow the two lamps to be dimmed independently.

Test yourself

1 Which one of the following is the correct symbol for an LDR?

    (a)   (b)   (c)

▲ **Figure 2.4**

2 Use words from the list below to label each of the components in the circuit in Figure 2.5.
  cell   resistor   fuse   lamp   switch   diode

3 Draw a circuit diagram to show a cell in series with an ammeter, variable resistor and lamp.

4 Draw a circuit diagram to show a cell lighting two lamps which are connected in parallel.

▲ **Figure 2.5**

○ Current and charge

Figure 2.6 shows a circuit diagram. In this circuit a cell provides a **potential difference** of 1.5 V, giving a current of 0.1 A in the circuit.

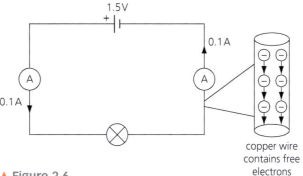

▲ Figure 2.6

KEY TERMS

⭐

Potential difference (p.d.) is a measure of the electrical work done by a cell (or other power supply) as charge flows round the circuit. Potential difference is measured in volts (V).

Electric current is a flow of electrical charge. The size of the electric current is the rate at which electrical charge flows round the circuit.

The **potential difference** (or p.d.) is a measure of the electrical work done by a cell (or other power supply) as charge flows round the circuit. The potential difference is measured in volts (V). Here the cell provides a potential difference of 1.5 V. (Remember the positive terminal of the cell is shown with the long line and the negative terminal with a shorter line.)

It is quite common to call the potential difference 'voltage'. However, you will find that the examination papers (and this book) will use the term potential difference.

In a metal the current is carried by electrons which are free to move. The electrons are repelled from the negative terminal of the cell and attracted towards the positive terminal.

In a circuit the direction of the **electric current** is always shown as the direction in which positive charge would flow – from the positive terminal of the battery to the negative terminal. Current was defined in this way before the electron was discovered, at a time when people did not understand how a wire carried a current. So in Figure 2.6 the direction of current is shown from positive to negative.

The amount of charge flowing round in the circuit is measured in **coulombs**, C. One coulomb of charge is equivalent to the charge on 6 billion billion electrons.

The unit of current is the **ampere**, A. This unit is often abbreviated to amp. Small currents can be measured in milliamps (mA).

1 mA = 0.001 A (10^{-3} A)

The current at all points of the circuit shown in Figure 2.6 is the same. So the two ammeters on either side of the lamp read the same current – in this case 0.1 A.

Current and the flow of charge are linked by the equation.

$$Q = It$$
charge flow = current × time
where charge is in coulombs, C
current is in amps, A
time is in seconds, s.

Example

In Figure 2.6 the current of 0.1 A flows for 30 minutes. How much charge flows round the circuit?

Answer

$Q = It$

$Q = 0.1 \times 1800$

$\quad = 180\,C$

Test yourself

5 State the unit of each of the following quantities:
   a) potential difference
   b) current
   c) charge.

6 In Figure 2.7 the current is 0.08 A at point X. What is the current at points Y and Z?

7 a) A charge of 3 C flows round a circuit in 2 seconds. Calculate the current.

   b) A torch battery delivers a current of 0.3 A for 20 minutes. Calculate the charge which flows round the circuit.

   c) A thunder cloud discharges 5 C of charge in 0.2 ms. Calculate the current.

   d) A mobile phone battery delivers a current of 0.1 mA for 30 minutes. Calculate the charge which flows through the battery in this time.

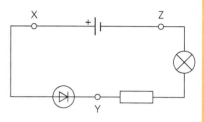

▲ Figure 2.7

Show you can...

Show you understand the nature of an electrical current, by explaining the relationship between a current and charge.

○ Controlling the current

You can change the size of the current in a circuit by changing the potential difference of the cell or battery, or by changing the components in the circuit.

> **TIP**
> In a series circuit, increasing the potential difference increases the current.

> **TIP**
> In a series circuit, increasing the resistance makes the current smaller. Resistance is a measure of a component's opposition to the current.

In Figure 2.8a) a current of 1 A flows. In Figure 2.8b) an extra cell has been added. Now the current is larger.

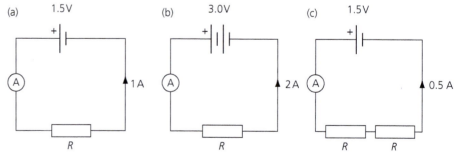

▲ Figure 2.8

> **KEY TERM**
> A **resistor** acts to limit the current in a circuit. When a resistor has a high resistance, the current is low.

In Figure 2.8c), the potential difference of the cell is 1.5 V but now the resistance has been increased by adding a second **resistor** to the circuit. This makes the current smaller.

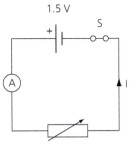

1.5 V

▲ Figure 2.9

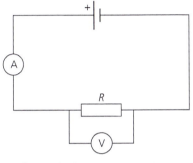

▲ Figure 2.10

▲ Figure 2.11 These ceramic discs have a very high resistance so that the power line is insulated from the metal pylon.

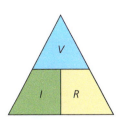

▲ Figure 2.12

In Figure 2.9 a variable resistor has replaced the fixed resistor in the circuit. You can control the size of the current by adjusting the resistor.

○ Ammeters and voltmeters

Figure 2.10 shows you how to set up a circuit using an ammeter and a voltmeter.

- The ammeter is set up in series with the resistor. The same current flows through the ammeter and the resistor.
- The voltmeter is placed in parallel with the resistor. The voltmeter measures the potential difference across (between the ends of) the resistor.
- The voltmeter only allows a very small current to flow through it, so it does not affect the current flowing around the circuit.

○ Resistance

The current through a component depends on two things:

- the resistance of the component
- the potential difference across the component.

The circuit in Figure 2.10 can be used to determine the value of the resistor in the circuit.

The current, potential difference or resistance can be calculated using the equation:

$$V = I\,R$$

potential difference = current × resistance

where potential difference is in volts, V
current is in amps, A
resistance is in ohms, Ω

Resistances are also measured in kilohms (kΩ) and megohms (MΩ).

$$1\,k\Omega = 1000\,\Omega\ (10^3\,\Omega) \quad 1\,M\Omega = 1\,000\,000\,\Omega\ (10^6\,\Omega)$$

It is very important that you can use this equation in each of its three forms. The triangle in Figure 2.12 is a useful way to remember it.

Example

A potential difference of 12V is applied across a resistor of 240Ω. Calculate the current through the resistor.

Answer

If you are in doubt about rearranging the formula on the examination paper, try to remember the triangle. Put your finger over the symbol you want to work out, in this case *I*, and we get:

$$I = \frac{V}{R}$$

$$= \frac{12}{240} = 0.05\,A$$

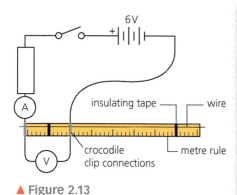

▲ Figure 2.13

Test yourself

8 You use a variable resistor to act as a dimmer for a torch lamp. How should you change the resistance to make the lamp brighter?

9 Copy and complete this table.

Electrical device	Potential difference across device in V	Current through device in A	Resistance of device in Ω
Resistor	1.5		20
Lamp	230	0.05	
Heater	230		23
LED		0.04	75
Electric car motor	72	48	

Required practical 3

Investigating how the resistance of a wire depends on the length of the wire

Method

1 Use electrical insulating tape to attach a 100 cm length of wire to a wooden metre rule. Very thin constantan wire is suitable with a diameter of 0.1 mm or less.

2 Connect the wire into the circuit as shown in Figure 2.13. Start by connecting the crocodile clips across 20 cm of the wire. Do not allow the current to rise higher than 1 A.

3 Draw a suitable table to record the length of the wire, the current through the wire, the potential difference across the wire and the calculated value for the resistance of the wire.

4 Close the switch then measure the current in the wire (I) and the potential difference (V) across the wire. Write these values and the length of the wire into your table.

5 Open the switch.

6 Use the equation $R = V/I$ to calculate the resistance of the wire.

7 Connect different lengths of wire into the circuit to obtain the data needed to plot a graph of resistance against length of wire.

If you have an ohmmeter you can measure the resistance of the wire directly; you do not need to calculate it.

Analysing the results

1 Plot a graph of resistance against the length of the wire.

2 Your graph should give a straight line going through the origin (0, 0). If it does, then you have shown that the resistance is **directly proportional** to the length of the wire.

Taking it further

Use the same circuit and a range of wires of different cross-sectional area to show that the resistance of a wire is **inversely proportional** to the area of the wire.

2 Set up the circuit shown in Figure 2.10 (page 41). Use the circuit to investigate the resistance of combinations of resistors in series and in parallel (page 46). First take measurements to calculate the value of a single fixed value resistor. Add a second resistor in series with the first resistor. Measure the p.d. across both resistors and the current through the resistors. Use these values to calculate the resistance of the two resistors in series. Repeat the experiment but with the two resistors connected in parallel. What do your results show happens to the resistance of a circuit when **a)** resistors are connected in series **b)** resistors are connected in parallel?

Questions

1 What was the dependent variable in this investigation?

2 What aspects of the investigation were important in trying to stop the wire from getting hot?

3 The width of the crocodile clips makes it difficult to measure the exact length of wire connected into the circuit. What type of error will this cause?

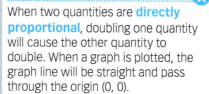

KEY TERMS

When two quantities are **directly proportional**, doubling one quantity will cause the other quantity to double. When a graph is plotted, the graph line will be straight and pass through the origin (0, 0).

When two quantities are **inversely proportional**, doubling one quantity will cause the other quantity to halve.

Current–potential difference characteristic graphs

An ohmic conductor

For some resistors, at constant temperature, the current through the resistor is proportional to the potential difference across it. A graph of current against potential difference gives a straight line. If the direction of the p.d. is reversed, the graph has the same shape. The resistance is the same when the current is reversed.

The resistor in this case is said to be **ohmic**.

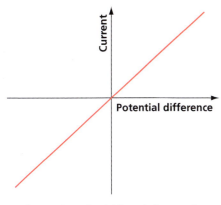

▲ **Figure 2.14** The *I–V* graph for a resistor or metal wire at constant temperature.

A filament lamp

The current–potential difference graph for a filament lamp does not give a straight line; the line curves away from the current axis (*y*-axis).

The current is not proportional to the applied potential difference. The lamp is a **non-ohmic** resistor.

As the current increases, the resistance gets larger. The temperature of the filament increases when the current increases. So we can conclude that the resistance of the filament increases as the temperature increases. Reversing the p.d. makes no difference to the way the resistance of the lamp changes. The resistance always increases when the temperature of the filament increases.

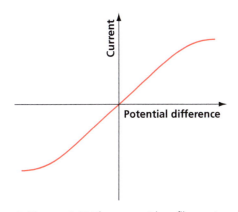

▲ **Figure 2.15** The current in a filament lamp does not increase in proportion to the potential difference.

> **KEY TERMS**
>
> The current flowing through an **ohmic** conductor is proportional to the potential difference across it. If the p.d. doubles, the current doubles. The resistance stays the same.
>
> The current flowing through a **non-ohmic** resistor is not proportional to the potential difference across it. The resistance changes as the current flowing through it changes.

A diode

A **diode** is a component that allows current to go in only one direction. For a forward potential difference, current starts to flow when the potential difference reaches about 0.7 V. When the potential difference is 'reversed', there is no current at all. The diode has a very high resistance in the reverse direction.

A **light-emitting diode** (LED) is a special type of diode that lights up when a current flows through it. This is useful because it allows an LED to be used as an indicator to show us that a small current is flowing.

Changing resistance

Some resistors change their resistance as they react to their surroundings. The resistance of a **thermistor** decreases as the temperature increases. You can control its temperature by putting it into a beaker of warm or cold water, as in Figure 2.17.

▲ **Figure 2.16** The resistance of a diode depends on the direction of the potential difference; the current only flows for a positive or forwards potential difference.

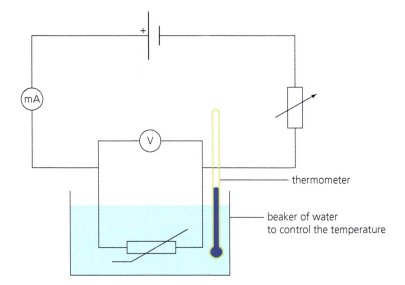

◀ **Figure 2.17** Investigating how the resistance of a thermistor changes with temperature.

By gently heating the water, the resistance of the thermistor can be found at different temperatures. You could use an ohmmeter to measure the resistance directly.

A thermistor can be used as the sensor in a temperature-operated circuit, such as a fire alarm. Some electronic thermometers use a thermistor to detect changes in temperature. The change in the resistance of a thermistor can be used to switch on (or off) other electrical circuits automatically.

The resistance of a **light-dependent resistor** (LDR) changes as the light intensity changes. In the dark the resistance is high but in bright light the resistance of an LDR is low. This is shown in Figure 2.19. A higher current flows through the resistor in bright light because the resistance is lower.

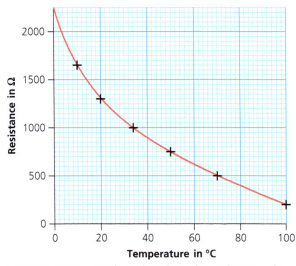

▲ **Figure 2.18** A graph to show how the resistance of a thermistor changes with temperature.

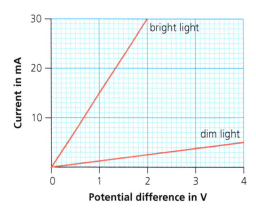

▲ **Figure 2.19** A current-potential difference graph for a light-dependent resistor in bright and dim light.

LDRs can be used as sensors in light-operated circuits, such as security lighting. The change in resistance of LDRs is used in digital cameras to control the total amount of light that enters the camera.

Required practical 4

Investigating the *I–V* characteristic of a circuit component

1 Set up the circuit shown in Figure 2.20. A suitable power supply to use is four 1.5V cells joined in series.

2 Connect a 6V filament lamp into the circuit where it says component.

3 Adjust the variable resistor to give a potential difference (p.d.) of 1V across the lamp.

4 Write the readings on the voltmeter (p.d.) and the ammeter (current) in a suitable table.

5 Adjust the variable resistor so that you can obtain a set of p.d. and current values. Write the new values in your table.

6 Reverse the connections to the power supply. The readings on the voltmeter and ammeter should now be negative. Obtain a new set of data with the p.d. increasing negatively.

You could leave the variable resistor out of the circuit and change the p.d. and current by simply connecting across one cell, then two cells, then three cells and lastly all four cells.

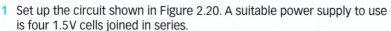

▲ Figure 2.20

Analysing the results

1 Plot a graph of current against potential difference (*I–V* characteristic graph). Draw the axes so that you can show all of the data, the positive values and the negative values.

2 You should notice that plotting the negative values for p.d. and current gives the same shape graph line as plotting the positive values.

3 If you used a variable resistor you would have been able to increase the p.d. using a smaller **interval** than if you simply connected across 1, 2, 3 then 4 cells. The advantage of using smaller intervals is that you can be more confident that the shape you draw for your graph line is correct.

> **KEY TERM**
>
> The **interval** is the difference between one value in a set of data and the next.

Taking it further

Replace the filament lamp with a low value resistor and then a diode. Obtain a set of p.d./current data for each component. Plot an *I–V* characteristic graph for each component. Remember to obtain the data needed to plot the negative part of the graph.

Questions

1 The *I–V* graph for a resistor (at constant temperature) is a straight line. Reversing the power supply does not change the shape of the graph line. What does this tell you about the resistance of the resistor and the direction of the current through the resistor?

2 How is the *I–V* graph for a diode different from the *I–V* graph for a filament lamp?

3 Why does having a small interval between values allow you to be more confident that the shape you draw for your graph line is correct?

4 If the p.d. were increased from 0V to 6V in 13 equal intervals, what would be the interval between p.d. values?

Test yourself

10 Use the graph in Figure 2.18 to find the resistance of the thermistor at temperatures of:
   a) 0 °C
   b) 40 °C
   c) 90 °C.
11 What is meant by an ohmic resistor?
12 Figure 2.21 shows the current–potential difference graph for a filament lamp.
   a) Calculate the resistance of the filament when the applied voltage is:
      i) 1 V
      ii) 3 V.
   b) What causes the resistance to change?
13 Figure 2.19 shows a current–potential difference graph for an LDR in dim and bright light. Calculate the resistance of the LDR in:
   a) bright light
   b) dim light.
14 a) Draw a circuit diagram to show a cell, a 1 kΩ resistor and an LED used to show that there is a current flowing through the resistor.
   b) Draw a circuit diagram to show how you would investigate the effect of light intensity on the resistance of an LDR.

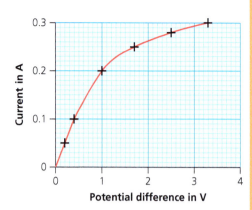

▲ Figure 2.21

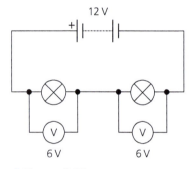

▲ Figure 2.22

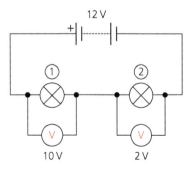

▲ Figure 2.23

Show you can...

Show that you understand about various types of resistors, by explaining how the following behave in a circuit:
a) an LDR
b) an LED
c) a thermistor.

○ Series and parallel circuits

Series circuits

In Figure 2.22 two identical lamps are connected in series with a 12 V battery. The p.d. of 12 V from the battery is shared equally so that each lamp has 6 V across it.

Having two lamps in the circuit rather than one increases the resistance, so the current decreases. The two lamps in series will not be as bright as a single lamp.

The potential differences do not always split equally. In Figure 2.23 lamp 1 has a larger resistance than lamp 2. There is a larger potential difference across lamp 1 than across lamp 2. As the current flows, lamp 1 transfers more energy to the surroundings than lamp 2. Therefore lamp 1 is brighter than lamp 2.

Series circuit rules

The rules for series circuits are as follows:

- there is the same current through each component
- the total potential difference of the power supply is shared between the components. So if there are just two components then;

$$V_{supply} = V_1 + V_2$$

- the total resistance of two components is the sum of the resistance of each component.

$$R_{total} = R_1 + R_2$$

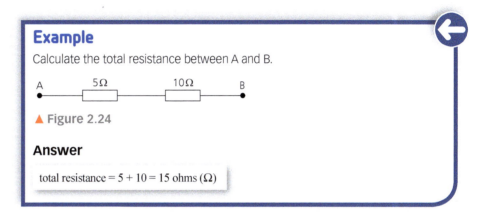

Example

Calculate the total resistance between A and B.

▲ Figure 2.24

Answer

total resistance = 5 + 10 = 15 ohms (Ω)

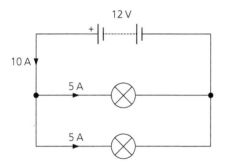

▲ Figure 2.25

▲ Figure 2.26 a) The cells are joined correctly – the separate p.d.s add to give 4.5 V.

▲ Figure 2.26 b) One cell is the wrong way round. The p.d. of this cell cancels out the p.d. of one of the other cells – the total p.d. is only 1.5 V.

Parallel circuits

In Figure 2.25 two identical lamps have been connected in parallel with the 12 V battery. Now there is a 12 V potential difference across each lamp, and the same current flows through each lamp.

When two lamps are joined in parallel, the total current in the circuit increases. So the combined resistance of the two lamps is less than either lamp by itself. The total current in the circuit is the sum of the currents through the two lamps.

Parallel circuit rules

The rules for parallel circuits are as follows:

- the potential difference across each component is the same
- the total current through the whole circuit is the sum of the currents through the separate components
- the total resistance of two resistors in parallel is less than the resistance of the smaller individual resistor.

Cells and batteries

A battery consists of two or more electrical cells. When cells are joined in series, the total potential difference of the battery is worked out by adding the separate potential differences together. This only works if the cells are joined facing the same way, positive (+) to negative (–) (see Figure 2.26).

Test yourself

15 Figure 2.27 shows a circuit with a cell, two ammeters and a resistor. What reading does the ammeter on the right–hand side give?

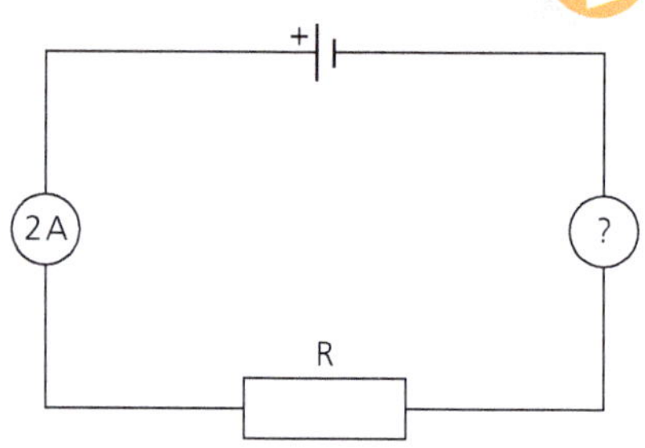

16 a) In Figure 2.28, what is the resistance between

   i) AB

   ii) CD?

   b) Which of the following correctly states the resistance between E and F?

30 Ω	more than 20 Ω
less than 10 Ω	between 20 Ω and 10 Ω

17 a) What is the potential difference of the cell in Figure 2.29a)?

   b) What is the potential difference across R_2 in Figure 2.29b)?

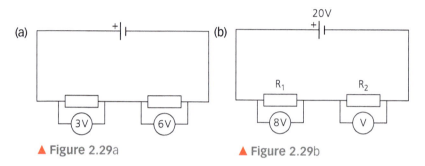

(a)

▲ Figure 2.29a

(b)

▲ Figure 2.29b

▲ Figure 2.27

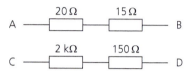

▲ Figure 2.28a

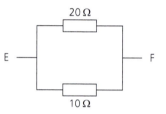

▲ Figure 2.28b

18 State the values of A_1, A_2, A_3 and A_4 in Figures 2.30a) and b).

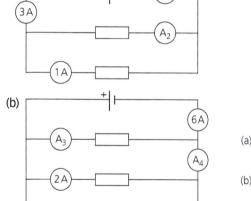

(a)

(b)

▲ Figure 2.30

(a) 6V 6V 6V

(b) 6V 6V 6V

(c) 12V 6V 9V

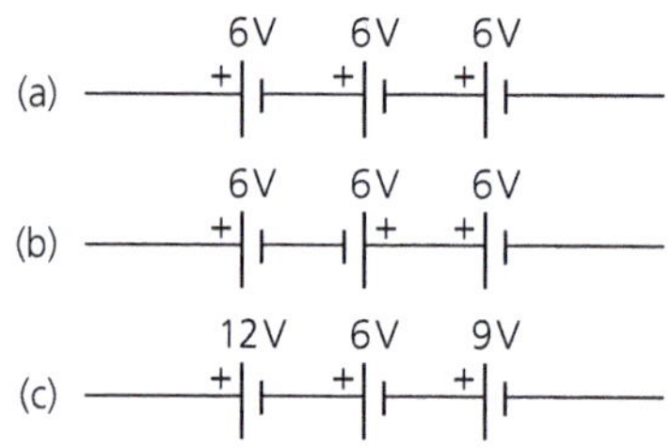

▲ Figure 2.31

19 Work out the potential difference of each cell combination shown in Figure 2.31.

○ Circuit calculations

You can use the series circuit rules to solve circuit problems. Two worked examples are given here.

Example

Use the information in Figure 2.32 to work out the resistance of the lamp.

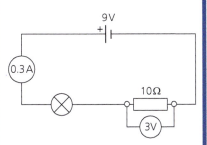

▲ Figure 2.32

Answer

The potential difference across the lamp is 9V – 3V = 6V

$$So \quad R = \frac{V}{I}$$
$$= \frac{6V}{0.3A}$$
$$= 20\,\Omega$$

Example

a) The light dependent resistor in Figure 2.33 is in bright light. Use the information in the diagram to work out its resistance.

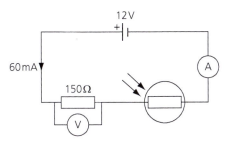

▲ Figure 2.33

b) Explain what happens to the voltmeter reading when the light intensity drops.

Answer

a) The total resistance of the series circuit can be calculated using:

$$R = \frac{V}{I}$$
$$= \frac{12}{0.06}$$
$$= 200\,\Omega$$

Therefore the LDR's resistance is:

$$200\,\Omega - 150\,\Omega = 50\,\Omega$$

OR you could work out the potential difference across the 150 Ω resistance:

$$V = I\,R$$
$$= 0.06 \times 150$$
$$= 9\,V$$

The potential difference across the LDR is:

$$12\,V - 9\,V = 3\,V$$

$$Then\ R = \frac{V}{I}$$
$$= \frac{3}{0.06}$$
$$= 50\,\Omega$$

b) When it gets dark the resistance of the LDR increases. So the current decreases. Therefore, the potential difference across the 150 Ω resistor drops, and the LDR gets a larger share of the potential difference.

Test yourself

20 a) Calculate the reading on the ammeter in Figure 2.34.

b) Now work out the potential difference, *V*, of the battery.

21 a) Calculate the reading on the ammeter in Figure 2.35.

b) Now work out the resistance *R*.

22 Work out the value of the resistance *R* in Figure 2.36.

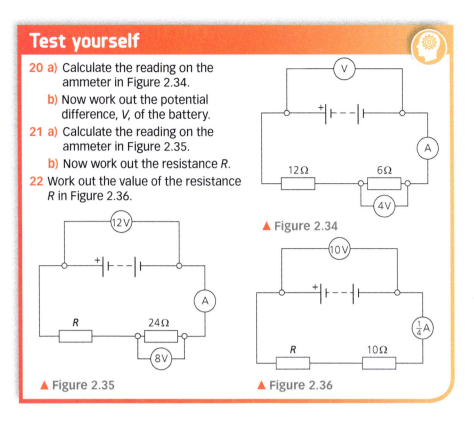

▲ Figure 2.34

▲ Figure 2.35

▲ Figure 2.36

▲ **Figure 2.37** Appliances in this kitchen are reliant on electricity to run

○ Domestic use and safety

In the home we rely on electricity for heating, lighting, cooking, washing and powering devices which we use for work or leisure.

Direct and alternating potential difference

A cell or battery provides potential difference.
A direct **potential difference** remains always in the same direction, and causes a current to flow in the same direction. This is a **direct current (d.c.)**.

If an alternating power supply is used in a circuit, the potential difference switches direction many times each second. This is an **alternating potential difference**, which causes the current to switch direction. So an **alternating current (a.c.)** is one that constantly changes direction, passing one way around a circuit and then the other.

Figure 2.38 shows graphs of how a.c. and d.c. power supplies change with time. The d.c. supply remains constant at 6V; the a.c. supply changes from positive to negative.

Note that the peak a.c. potential difference is a little higher than 6V. This is to make up for the time when the potential difference is close to zero. A 6V a.c. supply and a 6V d.c. supply will light the same lamp equally brightly.

Mains supply

The mains electricity is supplied by alternating current. In the UK it has a potential difference of about 230 V and a **frequency** of about 50 Hz.

KEY TERM

The unit of **frequency** is the hertz, Hz; 1 hertz means there is 1 cycle per second.

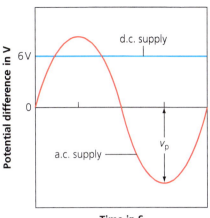

▲ Figure 2.38 Direct and alternating current supplies. V_p is the peak potential difference for the a.c. supply. You can use a cathode ray oscilloscope (CRO) to plot potential differences.

▲ Figure 2.39 A cable has 3 wires; neutral (blue), earth (green/yellow), live (brown).

A frequency of 50 Hz means that the cycle shown in Figure 2.38 repeats itself 50 times per second – or one cycle takes 1/50 of a second.

Test yourself

23 a) What is meant by the terms a.c. and d.c.?
   b) Explain the difference between direct and alternating potential difference.
24 In the USA the mains electricity supply is 115 V 60 Hz. Explain the difference between the mains electricity supply in the USA and the mains electricity supply in the UK.

Cables

We use many electrical appliances at home, which we connect to a wall socket using a three-core cable and plug. The wires inside the cables connect the appliance to the plug and have a cross-sectional area of 2.5 mm^2. These cables should carry no more than a 13 A current. Appliances such as showers and cookers need larger currents. These appliances are connected to the mains supply using thicker cables.

Live, neutral and earth wires

The insulation covering the three wires inside a cable is colour coded so we know which is which.

- **Live** wire – brown
- **Neutral** wire – blue
- **Earth** wire – green and yellow stripes

The live wire carries the alternating potential difference from the mains supply. The neutral wire completes the circuit. So the live and neutral wires carry the current to and from an electrical appliance. The earth wire is there to stop an appliance becoming live; it only carries a current if there is a fault.

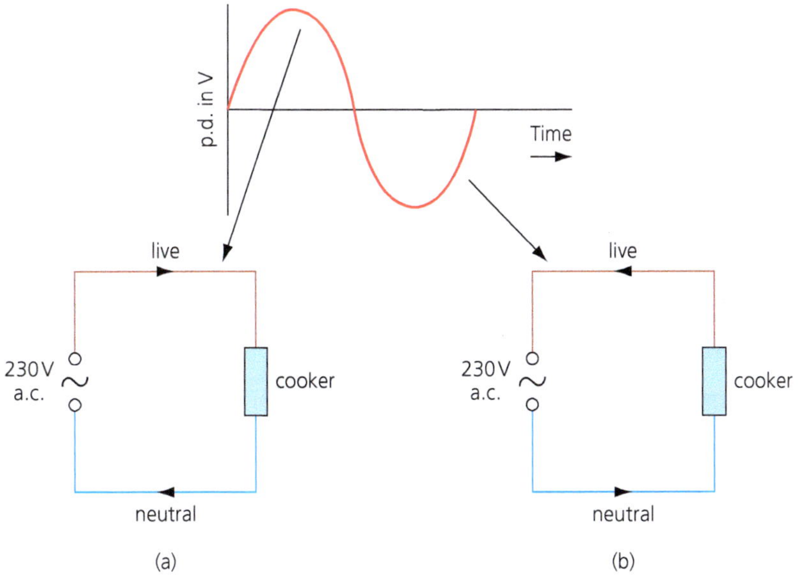

However, the three wires have different potentials.

- The earth wire is at 0 V.

- The potential difference between the live wire and earth (0 V) is about 230 V. A bare live wire is a hazard for us, even if it is not delivering a current to an appliance, that is switched off.

- The neutral wire is close to earth potential (0 V). So touching the neutral wire would not give us an electric shock.

In Figure 2.40, a 230 V a.c. supply provides current to a cooker. In diagram (a) the current goes one way round the circuit, then in diagram (b) the current is reversed. In each case energy is transferred to the cooker.

▲ Figure 2.40

Earthing

Any electrical appliance which has a metal case should be **earthed**. Figure 2.41 shows an electric toaster. It has been earthed by joining the earth wire from the three-core cable to the metal casing.

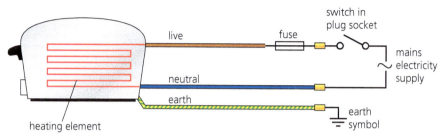

▲ Figure 2.41

The earth wire is a safety device to stop the appliance becoming live. So if a fault occurs and the live wire touches the toaster body, a person does not receive an electric shock.

> ## Test yourself
>
> **25 a)** What is meant by 'earthing' an appliance? Why are appliances earthed?
> **b)** Explain why touching the live wire of an appliance that is switched on is dangerous.

▲ **Figure 2.42** Why is this kettle earthed?

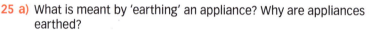

Power

○ Energy and charge

When charge passes through a resistor, the resistor gets hot. This is because electrons collide with the atoms in the resistor as they pass through it. The atoms increase their thermal energy store and vibrate faster, making the resistor hotter. The energy in the chemical store of the battery decreases as work is done to move the electrons round the circuit. The temperature of the resistor goes up and the resistor also heats up the surroundings.

The amount of energy transferred by electrical work depends on two factors:

- the potential difference, V, across the resistor
- how much charge, Q, flows through the resistor.

The energy transferred by electrical work can be calculated using this equation:

$$E = VQ$$

energy transferred = potential difference × charge

where energy is in joules, J
potential difference is in volts, V
charge is in coulombs, C

○ Power and energy

When an electricity company sends a bill to a household, the company is charging for the energy transferred. So knowing how much energy is transferred by an electrical appliance is important.

The energy transferred by an electrical appliance depends on two things:

- the power of the appliance
- the time the appliance is switched on.

So, the energy transferred by electrical work can also be calculated using this equation:

$$E = Pt$$

energy transferred = power × time

where energy is in joules, J
power is in watts, W
time is in seconds, s

By rearranging this equation, the power of an appliance can be calculated.

We can combine the two equations as follows.

$$P = \frac{VQ}{t}$$

but since

$$I = \frac{Q}{t}$$

$$P = VI$$

power = potential difference × current

where power is in watts, W
potential difference is in volts, V
current is in amps, A.

Since potential difference, $V = IR$, this equation can also be written as:

$$P = I^2 R$$

power = current squared × resistance

where power is in watts, W
current is in amps, A
resistance is in ohms, Ω.

Example

A 6V torch battery passes 250 C of charge through a lamp. How much energy has been transferred to the lamp?

Answer

$E = VQ$
$= 6 \times 250$
$= 1500\,\text{J}$

Example

A kettle has 0.5 kg of water in it. To heat the water from room temperature to boiling, 180 kJ of energy must be transferred. It takes 2 minutes for the kettle to boil. Calculate the power of the kettle.

Answer

$P = \frac{E}{t}$
$= \frac{180\,000}{120}$
$= 1500\,\text{W or } 1.5\,\text{kW}$

Remember: 180 kJ = 180 000 J
2 minutes = 120 s

In fact, the power of the kettle will be a little more than this, because some energy is wasted as the hot kettle will transfer some energy to heat up the surroundings.

Example 1

The information plate on a convection heater is marked as follows:

230V 50Hz 1800W

Calculate the current the heater draws from the mains supply and suggest which fuse, 3A or 13A, should be used in the plug.

Answer

$$P = VI$$
$$1800 = 230 \times I$$
$$I = \frac{1800}{230}$$
$$= 7.8\,A$$

So we must use a 13A fuse.

Example 2

A current of 4.7A passes through a 30 Ω resistor. Calculate the power transferred to the resistor.

Answer

$$P = I^2 R$$
$$= 4.7^2 \times 30$$
$$= 660\,W$$

TIP

You can remember the power equation in the form of this triangle. It helps you to rearrange the equation.

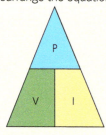

▲ Figure 2.43

Most electrical appliances have an information plate, which tells us the power of the appliance and the potential difference of the supply. From this you can work out the current the appliance will draw and the fuse needed for the appliance.

Test yourself

26 Calculate the power rating in watts of:
   a) a car starter motor that draws a current of 90A from a 12V supply
   b) a toaster that draws a current of 2.5A from a 230V supply
   c) a phone that draws a current of 0.0003A from a 3V battery.
27 Calculate the electrical work done by the supply in each of the following.
   a) A 12V battery supplies 200C of charge to a circuit.
   b) A current of 0.2A is drawn by a lamp from the mains 230V supply for 30 minutes.
   c) A current of 2mA is supplied by a 6V battery for 2 hours.
28 Calculate the current each of the appliances shown in the table draws from the mains supply, of 230V.

Appliance	Power
Lamp	11 W
TV	150 W
Hair dryer	480 W
Kettle	2200 W

○ National Grid

The electricity we use in our homes is generated in power stations. Our homes are often a long way from a power station and electricity is transmitted to us along overhead transmission cables. It is important to keep the current low, so that energy is not wasted in heating up the transmission cables.

▲ Figure 2.44

Figure 2.45 shows how we can reduce the current in the transmission cables.

- The generator in the power station sends a current of 1000 A at a potential difference of 25000 V into the National Grid.
- A device called a **transformer** steps the potential difference up to 400 000 V, but reduces the current to 62.5 A. (The operation of a transformer is explained in more detail in Chapter 7, page 236).
- With a current of 62.5 A, less energy is transferred into heating up the transmission cables than with a current of 1000 A.
- Near our homes, transformers step the potential difference down to a safe level of 230 V, so a larger current is available to use in the home.
- This makes the National Grid system an efficient way to transfer energy.

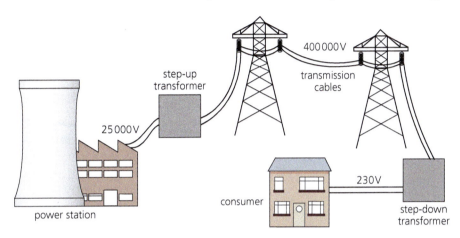

▲ **Figure 2.45** A step-up transformer increases the potential difference across the transmission cables. This allows the same power to be transmitted with a much lower current, reducing the energy wasted by heating up the cables.

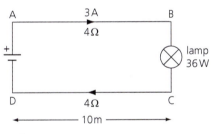

▲ Figure 2.46

Test yourself

29 In Figure 2.45 the resistance of the power line is 200 Ω.
   a) Calculate the power transferred in heating up the line when
      i) a current of 100 A passes through it
      ii) a current of 1000 A passes through it.
   b) Explain why transformers are used to reduce the current passing through the transmission cables of the National Grid.
30 A man decides to use a large cell in his house to supply current to light a 36 W lamp in his garden shed 10 m away (Figure 2.46). The resistances of the two wires to the shed are each 4 Ω.
   a) Calculate the power transferred in each wire.
   b) Calculate the fraction of the power delivered by the cell which is used to light the lamp.

Show you can...

Show you understand about the transmission of electricity by completing this task.

a) Why is electricity transmitted across the country at very high potential differences?
b) Why must we use only low potential differences in our homes?

Static electricity

When electric charge flows through a wire, a current flows. However, there are also many occasions when electrical charge is stationary (or static) on the surface of an object. This charge is called **static electricity**.

○ Making static

Static electricity can be produced by rubbing some insulating materials together. Negatively charged electrons are rubbed off one material and on to another. The material that gains electrons becomes negatively charged. The material that loses electrons becomes positively charged.

- When you comb your hair with a nylon comb, electrons are transferred to the comb. The comb becomes negatively charged. Your hair becomes positively charged.

- When you rub a perspex ruler with a duster, electrons are removed from the ruler and transferred to the duster. The ruler now has fewer electrons than protons so it has a positive charge. The duster has more electrons than protons, so the duster is negatively charged.

hair

▲ Figure 2.47

Sparks

The more you rub two insulators together, the more charge you can build up, which can produce a spark.

- You might have felt a shock after you have been walking on a carpet. By walking across the carpet you have built up a charge on yourself. When you touch an earthed metal object, a spark allows a current to flow. The charge stored on you flows quickly away.
- Lightning is a dramatic example of static electricity. Convection currents inside a thunder cloud cause ice crystals in the cloud to rub against each other. The bottom of a thunder cloud becomes negatively charged. The potential difference between the cloud and Earth is many millions of volts. The cloud is discharged by a flash (a large spark) of lightning.

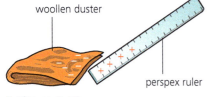

woollen duster

perspex ruler

▲ Figure 2.48

Energy transfers in lightning

- When a thunder cloud is charged, energy is stored in an electrostatic store.
- When the cloud discharges, electrical work is done when the charge flows from the cloud to Earth.
- Energy is transferred to the surroundings directly by heating (as the charge flows), but also by radiation and sound. We see lightning and hear thunder.
- The energy carried by the sound and light (and other radiation) is then transferred to the thermal store of the surroundings.

(a)
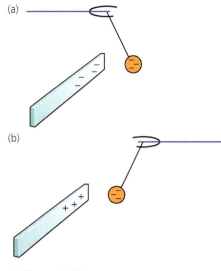

(b)

▲ Figure 2.49

▲ Figure 2.50 Lightning.

Forces on charges

Simple experiments show that charged objects exert a noticeable force on each other when they are brought close together. This is an example of a non-contact force.

● Two objects which carry the same type of charge repel each other.
● Two objects which carry different types of charge attract each other.

○ **Electric fields**

You are already familiar with two types of field.

● A gravitational field exerts a force on a mass. Earth's gravity keeps our feet on the ground.
● A magnetic field exerts a force on objects made of iron or steel.

In a similar way an electric field exerts a force on a charged object. Important points about an electric field are listed below:

● A charged object creates an electric field around itself.
● The electric field is strongest closest to the charged object.
● Further away from the charged object the field is weaker.
● The direction of the field is the direction of the force on a positive charge.
● The strength of the field can be increased by adding more charge to the object.

Figures 2.51a) and b) show the electric field close to a positively charged sphere and a negatively charged sphere.

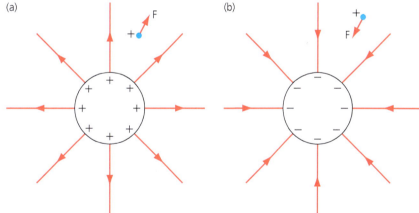
▲ Figure 2.51

● For the positively charged sphere, the direction of the field is away from the sphere.
● For the negatively charged sphere, the direction of the field is towards the sphere.
● As the distance increases from the sphere, the field lines get further apart. This shows that the field gets weaker further from the sphere.

Each diagram shows the direction of electric force on an isolated positive charge.

Test yourself

31 a) A plastic rod is rubbed with a cloth and the rod becomes positively charged. Explain what has happened to the rod.

  b) What charge does the cloth now have?

32 a) Figure 2.52 shows two small plastic balls. The balls are charged. What can you say about the charge on each ball?

  b) Draw a diagram to show the electrical forces acting on each ball.

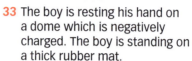

▲ Figure 2.52

33 The boy is resting his hand on a dome which is negatively charged. The boy is standing on a thick rubber mat.

  a) What is the charge on the boy's hand and hair?

  b) Why is the boy's hair sticking out?

  c) Why does the boy stand on a thick rubber mat?

34 A thunder cloud is charged to a potential difference of 100 MV. The cloud carries a charge of 5 C. A flash of lightning discharges the cloud in 0.002 s.

  a) Calculate the average current in the discharge.

  b) Calculate the energy transferred to the air during the lightning flash.

  c) Calculate the power produced by the lightning during the flash.

▲ Figure 2.53

Show you can...

Show you know about electrical charges by completing this task.

A girl rubs a balloon on her hair. After she has rubbed the balloon, her hair is attracted towards the balloon. Explain why.

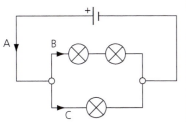

▲ Figure 2.54

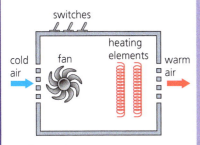

▲ Figure 2.55

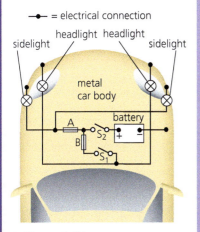

▲ Figure 2.56

Chapter review questions

1 Which current in Figure 2.54 is the largest, A, B or C? Which is the smallest?

2 Figure 2.55 shows a simplified picture of the inside of a small fan heater. The electrical wiring is not shown.

   Draw, using circuit symbols, a diagram to show how the heating element, fan and switches would be connected together so that:

   - when the mains is switched on, the fan comes on
   - both heating elements can be switched on independently
   - on hot days the heater can be used as a cooling fan.

3 Figure 2.56 shows a simplified circuit diagram for the front lights of a car. The metal body acts as a wire for the circuit.

   a) Which lights would not work if fuse A melted?

   b) Which switch operates the headlights?

   c) Can the sidelights be switched on without the headlights? Give a reason for your answer.

   d) If one headlight breaks, will the other still work? Give a reason for your answer.

   e) What change would you have to make to the circuit if the car's body was made of plastic?

4 A 10 Ω resistor is placed in series with a lamp, a switch and a 9 V battery.

   a) Draw the circuit diagram.

   b) When the switch is closed a current of 0.3 A flows. Calculate:

   i)   the power supplied by the battery

   ii) the power transferred to the resistor

   iii) the power transferred by the lamp.

5 In Figure 2.57 each cell has a potential difference of 1.5 V.

   a) What potential difference does the battery produce?

   b) State the reading on the voltmeter.

   c) Calculate the current through the ammeter.

   d) Calculate the resistance of i) the resistor, ii) the lamp.

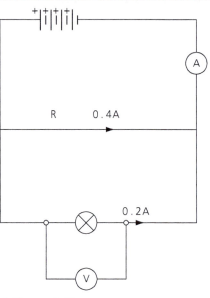

▲ Figure 2.57

6   In Figure 2.58 a 24 Ω resistor is placed in series with component X. Calculate the resistance of X.

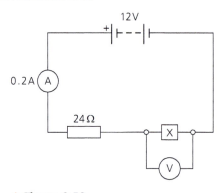

▲ Figure 2.58

7   A set of decorative lights has 115 identical lamps connected in series. Each lamp is designed to take a current of 0.05 A. The set of lamps is connected directly into the 230 V mains electricity supply.

a) What is the potential difference across one of the lamps?

b) Calculate the resistance of one of the lamps.

c) Calculate the resistance of all of the lamps in series.

d) Calculate the power of the set of lamps.

8   Explain why the resistance of a filament lamp increases as the current flowing through the lamp increases.

9   When the switch is closed in Figure 2.59, the lamp lights dimly at first. However, the lamp gets brighter slowly. Explain why.

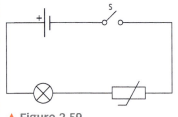

▲ Figure 2.59

Practice questions

1 Figure 2.60 shows a simple circuit.

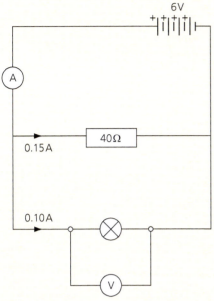

▲ Figure 2.60

a) The four cells are identical.

What is the potential difference of one cell? [1 mark]

b) State the reading on the voltmeter. [1 mark]

c) The current through the 40 Ω resistor is 0.15 A.

The current through the lamp is 0.10 A.

What is the reading on the ammeter? [1 mark]

d) Use the correct answer from the box to copy and complete the sentence.

less than	equal to	greater than

The lamp has a resistance _____ 40 Ω

Give a reason for your answer. [2 marks]

2 A student rubs a nylon comb against the sleeve of his woollen jumper.

a) Use the correct words from the box to copy and complete the sentence.

arm	jumper	negatively	positively

The comb becomes _____
charged because electrons move from the student's
_____ onto the comb. [2 marks]

b) The statement in the box is wrong.

Objects carrying opposite charge will repel each other.

Change one word to make the statement correct.

Write down the corrected statement. [1 mark]

c)
i) A tall building is fitted with a lightning conductor. If the building is struck by lightning the charge flows through the lightning conductor to Earth.

From which of the following materials should the lightning conductor be made?

copper	plastic	rubber

Give the reason for your answer. [2 marks]

ii) Use the correct answer from the box to copy and complete the sentence.

less than	equal to	greater than

The resistance of the building is
_____ the resistance of the
lightning conductor. [1 mark]

iii) Which of the following words is used to describe the flow of charge through a lightning conductor?

current	power	resistance

[1 mark]

3 Figure 2.61 shows a simple circuit. The circuit includes an LDR.

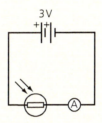

▲ Figure 2.61

a) How does the resistance of an LDR change with changing light intensity? [1 mark]

b) Figure 2.62 shows how the reading on the ammeter changes with light intensity.

▲ Figure 2.62

i) What is the current in the circuit when the light intensity is equal to the value marked 'X'? [1 mark]

ii) Calculate the resistance of the LDR when the light intensity is equal to the value 'X'. Give the unit. [2 marks]

iii) Suggest a practical use for this circuit. [1 mark]

c) Figure 2.63 shows the current–potential difference graph for an LDR in a dark room.

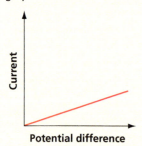

▲ Figure 2.63

Copy the graph and draw a second line to show how the current would change with potential difference if the LDR were in bright sunlight. [1 mark]

4 An electric iron has been wired without an earth connection. After years of use the live wire becomes loose and touches the metal part of the iron.

a) A man touches the iron and receives an electric shock. Sketch a diagram to show the path taken by the current. [1 mark]

b) The mains potential difference is 230 V. The man's resistance is 46 kΩ.

Calculate the current that passes through the man. [3 marks]

5 Figure 2.64 shows three resistors connected to a 12 V battery.

a) Calculate the currents through the ammeters A_1 and A_2. [2 marks]

b) Which resistor has the greater value, R_1 or R_2. Give a reason for your answer. [2 marks]

c) Calculate the resistance R_1. [3 marks]

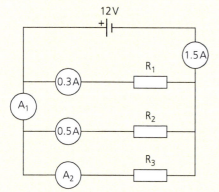

▲ Figure 2.64

6 A filament lamp is connected to a 12 V battery. The current through the lamp is recorded by a data logger when the lamp is switched on. Figure 2.65 shows how the current changes just after the lamp is switched on.

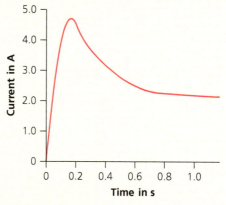

▲ Figure 2.65

a) Describe how the current changes just after the lamp is switched on. [2 marks]

b) Use the graph to determine:

i) the maximum current

ii) the current after 1 second. [2 marks]

c) The resistance of the filament increases as it gets hotter. Use this information to explain the shape of the graph. [3 marks]

d) Use the graph to calculate the power of the lamp when it is working at its steady temperature. [3 marks]

7 Figure 2.66 shows a circuit diagram which includes a diode. Figure 2.67 shows how the current through the diode varies with the potential difference across the diode.

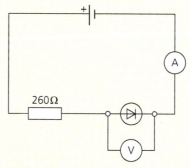

▲ Figure 2.66

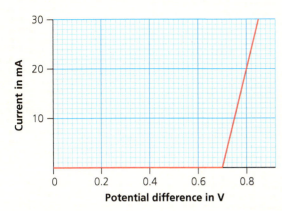

▲ Figure 2.67

a) A student sets up the circuit and measures the current through the diode as 20 mA. Use Figure 2.67 to determine the potential difference across the diode for this current. [1 mark]

b) Calculate the potential difference across the 260 Ω resistor when the current through the resistor is 20 mA. [3 marks]

c) Calculate the potential difference of the cell. [1 mark]

8 Figure 2.68 shows a circuit which includes a fixed resistor of 750 Ω and a component, X. The resistance of X changes with temperature. Figure 2.69 shows how the resistance of X changes with temperature.

a) What is component X? [1 mark]

b) At what temperature does X have a resistance of 250 Ω? [1 mark]

c) Calculate the current flowing through the circuit when X has a resistance of 250 Ω. [3 marks]

The component is now placed in a beaker of water and warmed from 20°C to 100°C.

d) Describe how the reading on the voltmeter changes as the water warms from 20°C to 100°C. Give reasons for your answer. [3 marks]

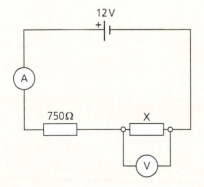

▲ Figure 2.68

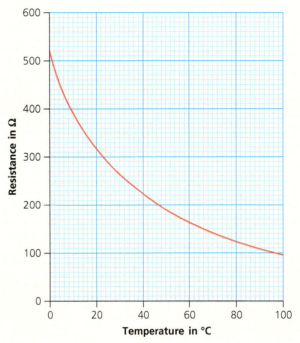

▲ Figure 2.69

9 a) The table shows the current in three different electrical appliances when connected to a 230 V a.c. supply.

Appliance	Current in A
Kettle	11.5
Lamp	0.05
Toaster	4.2

i) Which appliance has the greatest resistance? How does the data show this? [2 marks]

ii) The lamp is connected to the mains supply using a thin, twin-cored cable, consisting of live and neutral connections. State two reasons why this cable should not be used to connect the kettle to the mains supply. [2 marks]

b) Calculate the power rating of the kettle when it is operated from the 230 V a.c. mains supply. [3 marks]

The kettle is taken to the USA where the mains supply has a potential difference of 115 V.

c) i) Calculate the current flowing through the kettle when it is connected to a 115 V mains supply. [3 marks]

ii) The kettle is filled with water. The water takes 90 s to boil when working from the 230 V supply. Calculate how the time it takes to boil changes when the kettle operates on the 115 V supply. [3 marks]

Working scientifically

Units and calibration

International System of Units (SI)

The three thermometers in Figure 2.70 are all measuring the same temperature but each one gives a different reading. This is because each thermometer has been calibrated using a different scale of units.

Scientists around the world have an agreed set of units that are used to measure quantities such as mass, time, current and temperature. These are known as the SI units. The units used in this book are SI units.

1 The table lists three quantities and the SI unit for that quantity.

Quantity	Unit	Symbol
Current	ampere	A
Resistance	ohm	Ω
Energy	joule	J

Draw your own table and list all of the quantities and their SI units found in this chapter.

2 Suggest why it is important that scientists around the world measure quantities using the same system of units.

Powers of ten

Some measurements that we make may be very small or very large. An electric current may be very small because the resistance of the circuit is very large. In this case, we may measure the current in milliamps and the resistance in kilohms.

The table lists the prefixes and powers of ten that you need to be able to use.

Prefix	Symbol	Power of ten
tera	T	10^{12}
giga	G	10^{9}
mega	M	10^{6}
kilo	k	10^{3}
centi	c	10^{-2}
milli	m	10^{-3}
micro	μ	10^{-6}
nano	n	10^{-9}

Calibrating a voltmeter to measure temperature

Zach has designed the circuit shown in Figure 2.71 to measure temperature. The circuit includes a thermistor. We know that the resistance of a thermistor changes with temperature, so the thermistor is being used as the temperature sensor.

The reading on the voltmeter changes as the resistance of the thermistor changes. This means that the voltmeter can be used to measure temperature, but first it must be calibrated.

KEY TERM

Calibrate means to mark a scale onto a measuring instrument so that you can give a value to a measured quantity.

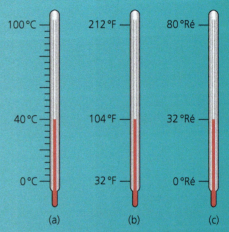

(a)   (b)   (c)

▲ Figure 2.70 a) This thermometer is calibrated in degrees Celsius (°C). This scale is the one you use in science.

b) This thermometer is calibrated in degrees Fahrenheit (°F). This scale is often used in UK weather forecasts.

c) This thermometer is calibrated in degrees Réaumur (°Ré). This is an old scale that is not used anymore.

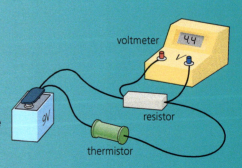

▲ Figure 2.71

To calibrate the voltmeter to measure temperature Zach put the thermistor into a beaker of ice cold water at exactly 0 °C. Zach then heated the water. Using an accurate thermometer, Zach measured and recorded different water temperatures and the reading of the voltmeter at those temperatures. Zach then drew the calibration graph shown in Figure 2.72.

3 Explain how Zach can use the calibration graph to convert a voltmeter reading into a temperature value.

4 Explain why Zach could not have drawn an accurate calibration graph if he had only put the thermistor into melting ice and boiling water.

5 Estimate the reading on the voltmeter if the thermistor were to be placed inside an oven at 120 °C. (The rest of the circuit is outside the oven). To do this you must assume the pattern shown in Figure 2.72 continues – this is called **extrapolating** the results. Extrapolating results is easiest if the pattern is a straight line.

6 Explain how the resolution of the voltmeter as a thermometer changes as the temperature of the thermistor increases above 60 °C.

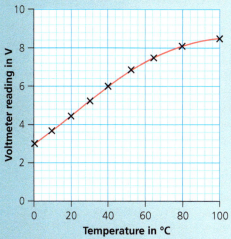

▲ Figure 2.72

KEY TERM

Extrapolation is to make an estimate (or prediction) assuming that an existing trend or pattern continues to apply in an unknown situation.

3 Particle model of matter

In the nineteenth century scientists proposed the existence of atoms to explain some of their observations – for example the random motion of small particles of pollen, seen floating on water. By the early twentieth century, we had worked out the size of atoms, but in the twenty first century we can use a scanning tunnelling microscope to form an image of atoms in solids. In the photograph, each green sphere is a silicon atom. The image is generated by a computer, based on the tiny electron current detected by a probe near the surface of the silicon sample.

Specification coverage

This chapter covers specification point: 4.3 Particle model of matter. It covers 4.3.1 Changes of state and the particle model, 4.3.2 Internal energy and energy transfers, and 4.3.3 Particle model and pressure.

Previously you could have learned

> Matter is made up of atoms and molecules.
> There are three states of matter: solid, liquid, gas.
> In a solid, atoms are packed together in regular patterns.
> In a liquid, atoms are in contact and are able to flow past each other.
> In a gas, atoms or molecules are spread out and are free to move around.
> The density of a substance measures the mass of that substance in a given volume. Solids and liquids are usually denser than gases.

Test yourself on prior knowledge

1 Name the three states of matter.
2 Explain why 1 m³ of gas (at atmospheric pressure) has less mass than 1 m³ of a solid material.
3 Why does 1 kg of water have a smaller volume than 1 kg of steam?

Prior knowledge

Density

A tree with a mass of 1000 kg obviously has a greater mass than a steel nail with a mass of 0.01 kg. However, sometimes we hear people say: 'steel is heavier than wood.' What they mean is that a piece of steel has a greater mass than a piece of wood with the same volume. Steel has a greater **density** than wood.

The density of a material is defined by the equation:

$$\rho = \frac{m}{V}$$

$$\text{density} = \frac{\text{mass}}{\text{volume}}$$

where density, ρ, is in kilograms per metre cubed, kg/m³

mass, m, is in kilograms, kg

volume, V, is in metres cubed, m³.

Example

Aluminium has a density of 2700 kg/m³. Calculate the volume of 135 g of aluminium.

Answer

$$\rho = \frac{m}{V}$$

$$2700 = \frac{0.135}{V}$$

$$V = \frac{0.135}{2700}$$

$$= 0.00005 \, \text{m}^3$$

$$= 5 \times 10^{-5} \, \text{m}^3$$

Remember: always make sure to work in kg and m³.

An investigation to calculate the density of liquids and solids.

To calculate the density of a substance, the mass and volume of the substance must be measured. Density can then be calculated using the equation:

$$\text{density} = \frac{\text{mass}}{\text{volume}}$$

The density of a liquid

Method

1   Draw a table to write your results in. Make sure each column in the table has a heading that includes the quantity and the unit.

2   Use an electronic balance to measure the mass of an empty 100 millilitre (ml) measuring cylinder.

3   Take the measuring cylinder off the balance then carefully pour 20 ml of water into the measuring cylinder.

4   Measure the new mass of the measuring cylinder and water, then calculate the mass of water in the measuring cylinder.

5   Add another 20 ml of water to the measuring cylinder then measure the new mass. Repeat this by adding another 20 ml of water to the measuring cylinder.

6   You now have three sets of results for different masses and volume of water.

(a)

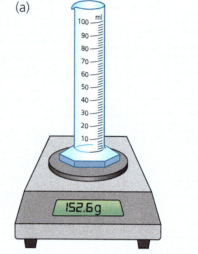

(b)

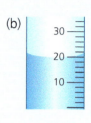

▲ **Figure 3.1** a) Use an electronic balance with a resolution of at least 0.1 g. b) The surface of the water is curved (the meniscus). Measure the volume at the lowest point of the curve.

Analysing the results

1   Use each set of results to calculate a value for the density of water. Calculate the mean (average) value.

Note – a volume of 1 ml is the same as 1 centimetre cubed (cm^3).

2   If you have mass in grams (g) and volume in cm^3 then density will be in g/cm^3. Change the density to kg/m^3 by multiplying by 1000.

Taking it further

Olive oil and vinegar are often used to make a salad dressing. When left in a bottle they eventually separate out. The one with the highest density will sink to the bottom. Which one would you predict would go to the bottom of the bottle, the oil or the vinegar?

Measure the densities of olive oil and of vinegar and see if your prediction was right.

Questions

1   Why is it better to measure a large volume of liquid rather than a small volume?

2   What is meant by a resolution of 0.1 g?

The density of a regular solid

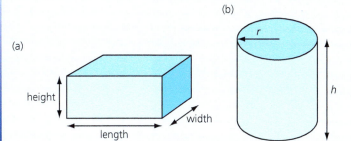

(a)

(b)

▲ **Figure 3.2** a) The volume of a cuboid = length × width × height.
b) The volume of a cylinder = $\pi r^2 h$.

Method

1 Measure the dimensions of the solid. If it is a cuboid this will be the length, width and height. Measure each dimension several times in different places. If the measurements for, say, the length are different, calculate a mean value. Then calculate the volume of the solid.

2 Measure the mass of the solid.

Analysing the results

1 Calculate the density of the solid.

2 If you know the type of material your solid is, look up its true density. Calculate the difference between your value and the true value. Do you think your value is accurate?

Taking it further

Describe how to measure the density of a sheet of aluminium cooking foil.

Questions

1 Measuring the length of a cuboid three times may give three slightly different values. Suggest why.

2 Describe how you can measure the thickness of paper using an ordinary 30 cm ruler.

The density of an irregularly shaped solid

Method

1 Make sure the object to be used fits easily into the measuring cylinder.

2 Measure the mass of the object.

3 Put enough water into the measuring cylinder to submerge the object. Measure the volume of water in the measuring cylinder.

4 Tilt the measuring cylinder and slide the object in.

5 Measure the new position of the water surface in the measuring cylinder. Then calculate the volume of the object.

Analysing the results

1 Calculate the density of the object.

Taking it further

Use several different shaped objects all of the same material. This could simply be five or six stones. Measure the mass and volume of each object. Plot a graph of mass against volume. Your graph may look like Figure 3.4.

▲ **Figure 3.3** The volume of the object = volume of water displaced.

Use the graph to calculate the density of the objects.

Your graph may include an **anomalous** data point. This could be due to a big measurement error but more likely one of the objects has a different density. If so then the object must be a different material.

Questions

1 Why should a graph of mass against volume go through the origin?

2 How does plotting a graph allow you to identify anomalous data?

KEY TERM ⭐

An **anomalous** result is one that does not fit the expected pattern.

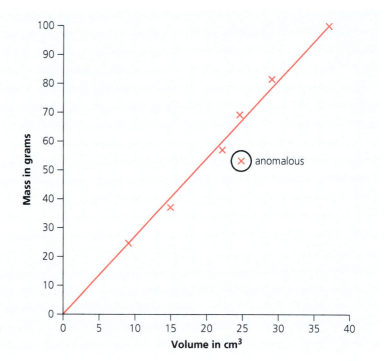

▶ **Figure 3.4** If the line misses the origin, there may have been an error in the measurements.

Test yourself

1 When answering an exam question a student wrote the following.

'A cork floats on water because it is lighter than water. A stone sinks because it is too heavy to float on water.'

Correct the mistakes in the student's answer.

2 A student wants to determine the density of a cuboid of material. He takes these measurements.

- Mass of the cuboid = 173.2 g
- Length = 10.1 cm; Width = 4.8 cm; Height = 1.3 cm

Calculate the density of the material in kg/m³.

3 A geologist needs to determine the density of a rock. First she weighs the rock and calculates that its mass is 90 g. Next she measures the volume of the rock by immersing it in water.

a) i) Use the information in Figure 3.5 to determine the volume of the rock in ml.

   ii) Express this volume in m³. [1 ml = 1 cm³ = 10^{-6} m³]

b) Calculate the density of the rock in kg/m³.

4 Copy the table and fill in the gaps.

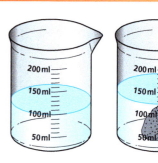

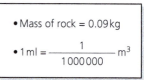

- Mass of rock = 0.09 kg
- $1\,ml = \dfrac{1}{1\,000\,000}\,m^3$

▲ **Figure 3.5**

Material	Volume in m³	Mass in kg	Density in kg/m³
Water	3	3000	
Alcohol		3200	800
Titanium	0.5		4500
Cork		0.2	200
Gold	0.02	390	

Show you can...

Show that you understand the definition of density by describing an experiment to calculate the density of a liquid. Explain what measurements you will take and the calculations you will do.

Solid, liquids and gases

Ice, water and steam are three different **states** of the same substance. We call these three states solid, liquid and gas.

- **Solid**: In a solid the atoms (or molecules) are packed in a regular structure. The atoms cannot move out of their fixed position, but they can vibrate. The atoms are held close together by strong forces. So it is difficult to change the shape of a solid.
- **Liquid:** The atoms (or molecules) in a liquid are also close together. The forces between the atoms keep them in contact, but atoms can move from one place to another. A liquid can flow and change shape to fit any container. Because the atoms are close together, it is very difficult to compress a liquid.
- **Gas:** In a gas the atoms (or molecules) are separated by relatively large distances. The forces between the atoms are very small. The atoms in a gas are in a constant random motion. A gas can expand to fill any volume.

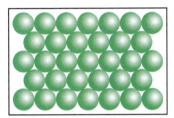 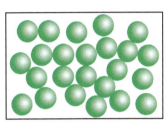 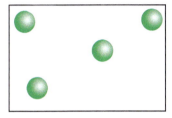

▲ **Figure 3.6** The particle arrangement in a solid, a liquid and a gas.

○ Density of solids, liquids and gases

The densities of solids and liquids are usually much higher than the density of gases. In solids and liquids, the atoms are closely packed together, so there is a lot of mass in a small volume. In gases the atoms are much further apart, so there is less mass in the same small volume.

The densities of some solids, liquids and gases. (The gases are at room temperature and pressure.)

Material	Density in kg/m³
Lead (solid)	11 400
Glass (solid)	2 500
Water (liquid)	1 000
Lithium (solid)	500
Cork (solid)	200
Air (gas)	1.3
Hydrogen (gas)	0.09

Test yourself

5 Draw diagrams to show the arrangement of the particles in each of the three states of matter.

6 Why are gases less dense than liquids and solids?

- Lead has a much higher density than lithium because the atoms of lead have a much greater mass than lithium atoms. (The lead atoms are only slightly larger than lithium atoms.)
- The density of air is larger than the density of hydrogen, because the nitrogen and oxygen molecules in the air have a greater mass than hydrogen molecules.

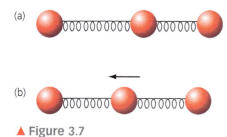

(a)

(b)

▲ Figure 3.7

○ Internal energy

Energy is stored inside a system by the particles (atoms or molecules) that make up that system. This is called **internal energy**. Internal energy is the total kinetic and potential energy of all the particles that make up the system.

We can use a model of several balls and springs to help us understand the nature of internal energy in a solid. The balls represent the atoms and the springs represent the forces or 'bonds' that keep the atom in place.

In Figure 3.7a) the middle atom has been displaced to the right. Now potential energy is stored in the stretched bond. In Figure 3.7b) the atom has kinetic energy as it moves to the left.

Heating

Heating changes the energy stored within a system by increasing the energy of the particles that make up the system.

● Heating can increase the temperature of a system. For example, when a gas is heated, the atoms (or molecules) move faster and the kinetic energy of the atoms increases.
● Heating a system can also cause a change of state; for example, when a solid melts to become a liquid. Usually when a solid melts, there is a small increase in volume, as the solid turns to liquid. The atoms increase their separation and there is an increase in the potential energy stored. So the internal energy increases.

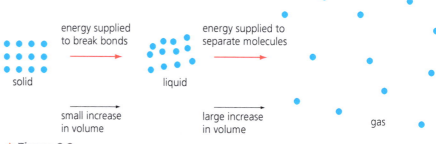

▲ Figure 3.8

Changes of state

We use these terms to describe changes of state.

● **Melting** occurs when a solid turns to a liquid. The internal energy of the system increases.
● **Freezing** occurs when a liquid turns to a solid. The internal energy of the system decreases.
● **Boiling** or **evaporation** occurs when a liquid turns to a gas. The internal energy of the system increases.
● **Condensation** occurs when a gas turns to a liquid. The internal energy of the system decreases.
● **Sublimation** occurs when a solid turns directly into a gas. The internal energy of the system increases. Sublimation is rare. An example is carbon dioxide (CO_2): solid CO_2 (dry ice) turns directly into the gas CO_2 missing out the liquid state at normal atmospheric pressure.

> **TIPS**
> Evaporation of a liquid takes place at any temperature. Your wet washing dries as water evaporates.
>
> Boiling occurs at a liquid's boiling point. At this temperature, bubbles form in the liquid.

A change of state of a substance is a **physical change**. The change does not produce a new substance and the process can be reversed. For example, a cube of ice from the freezer can be allowed to melt into water. The water can be put back into its container and then into the freezer. The water will freeze back into ice. No matter what its state, water or ice, the mass is the same. So, when a substance changes state, the mass is conserved. This is because the total number of particles (atoms or molecules) stays the same.

Test yourself

7 a) What is meant by a change of state of a substance?
  b) Give two examples of changes of state.
8 Which of the following changes are physical changes?
  * Melting snow
  * Burning a matchstick
  * Breaking a matchstick
  * Boiling an egg
  * Mixing salt and sugar together.
9 a) What happens to the internal energy of a system when the system is heated?
  b) How is it possible to heat a system without the temperature of the system increasing?

Specific heat capacity

When the temperature of a system is increased by supplying energy to it, the increase in temperature depends on:

* the mass of the substance heated
* what the substance is made of
* the energy put into the system.

Water needs much more energy to increase its temperature by 1 °C than the same mass of concrete. This also means that when water cools by 1 °C, it gives out more energy than the same mass of concrete cooling by 1 °C.

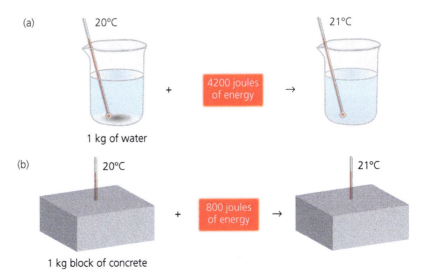

▶ **Figure 3.9** a) 4200 joules of energy are needed to increase the temperature of 1 kg of water by 1 °C. b) 800 joules of energy are needed to increase the temperature of 1 kg of concrete by 1 °C.

Example

A domestic hot water tank contains 200 kg of water. How much energy is required to warm the tank from 15 °C to 45 °C?

Answer

Temperature rise
= 45 °C − 15 °C = 30 °C

Energy supplied = increase in thermal energy of the water

$\Delta E = m\, c\, \Delta\theta$
$= 200 \times 4200 \times 30$
$= 25\,200\,000\, \text{J}$
$= 25.2\, \text{MJ}$

We say that the **specific heat capacity** of water is 4200 joules per kilogram per degree Celsius (J/kg °C).

The specific heat capacity of a substance is the amount of energy required to raise the temperature of one kilogram of the substance by one degree Celsius.

To calculate the change in thermal energy in a substance we use the equation:

$$\Delta E = mc\, \Delta\theta$$

$$\text{change in thermal energy} = \text{mass} \times \text{specific heat capacity} \times \text{temperature change}$$

Where change in thermal energy is in joules, J

mass is in kilograms, kg

specific heat capacity is in joules per kilogram per degree Celsius, J/kg °C

temperature change is in degrees Celsius, °C.

The table gives some examples of specific heat capacities for various substances at 20 °C.

Examples of specific heat capacities (c)

Substance	Specific heat capacity in J/kg °C
Water	4200
Alcohol	2400
Ice	2100
Dry air	1000
Aluminium	880
Concrete	800
Glass	630
Steel	450
Copper	380
Lead	160

◯ The specific heat capacity of water

Water has a very high specific heat capacity. This means that 1 kg of water requires a lot of energy to heat it up and a lot of energy must be transferred from the water when it cools down. This high specific heat capacity is very important.

- We are made mostly of water. A high specific heat capacity of water means that our body temperature does not increase too much when we take exercise or cool too quickly when we go outside on a cold day.
- Water is used for keeping many homes warm. A house central heating system pumps hot water around the house. The hot water transfers a lot of energy as it flows through radiators. If water had a low specific heat capacity, water would cool down before it got to some of the radiators in your house.

Test yourself

In these questions you will need to refer to the information in the table on page 74.

10 a) A night storage heater contains 60 kg of concrete. A heater embedded in the concrete heats the concrete up from 12 °C to 37 °C. How much energy is transferred to the concrete?

b) A heater supplies 4180 J to a block of copper of mass 0.5 kg. Calculate the temperature rise of the block.

c) A heater supplies 21 120 J to a block of aluminium. The temperature of the block rises from 18 °C to 34 °C. Calculate the mass of the block.

11 a) In Figure 3.10 a block of tin is heated from a temperature of 20 °C to 65 °C. The mass of the block is 1.2 kg. Use the reading on the joulemeter to calculate the specific heat capacity of tin.

b) Give two reasons why the specific heat capacity calculated in part a) is likely to be inaccurate.

12 An electric kettle has a power rating of 2.0 kW. The kettle is filled with 0.75 kg of water.

a) Calculate the energy required to warm the water from 20 °C to 100 °C.

b) Calculate how long it takes the kettle to bring the water to the boil at 100 °C from 20 °C.

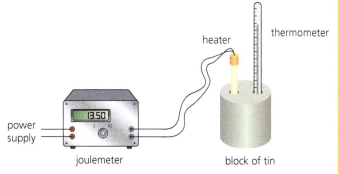

▲ **Figure 3.10**

TIP ✔

Specific heat capacity has units of J/kg °C. You will be expected to know this for your examination.

◯ Latent heat

When you heat a pan of water on the cooker, energy is transferred to the water and the temperature of the water increases. After a while the water begins to boil, and the temperature of the water stays constant at 100 °C. Yet, the cooker is still supplying energy to the water at the same rate. So where is the energy transferred to now? The answer is that energy is transferred into the internal energy of the steam. The molecules in steam at 100 °C have more internal energy than the same molecules of water at 100 °C.

The energy needed for 1 kg of a substance to change state is called **specific latent heat**.

The specific latent heat of a substance is the amount of energy required to change the state of one kilogram of the substance with no change in temperature.

The energy required to change the state of a substance can be calculated using this equation:

$$E = mL$$

energy required = mass × specific latent heat

where energy is in joules, J
       mass is in kilograms, kg
       specific latent heat is in Joules per kilogram, J/kg.

TIP

When a change of state happens, the energy supplied changes the energy stored (the internal energy) but not the temperature.

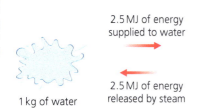

2.5 MJ of energy supplied to water

2.5 MJ of energy released by steam

1 kg of water at 100 °C

1 kg of steam at 100 °C

▲ Figure 3.11

Matter has three states: solid, liquid and gas. So a substance has two specific latent heats.

● The specific latent heat of **fusion** is the energy required to change 1 kg of a solid into 1 kg of a liquid at the same temperature.
● The specific latent heat of **vaporisation** is the energy required to change 1 kg of a liquid into 1 kg of a gas at the same temperature.

Melting and freezing

When a substance melts, energy must be supplied to the substance.

When a substance freezes (or solidifies), energy is transferred from the substance to the surroundings.

Vaporising and condensing

When a substance vaporises, energy is supplied to the substance to turn it from a liquid into a gas.

When a substance condenses, energy is transferred from the substance as it changes from a gas into a liquid.

○ **Measuring latent heat**

Figure 3.12 shows how you can calculate the latent heat of vaporisation of water.

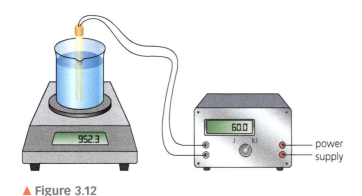

power supply

▲ Figure 3.12

● A beaker of water is placed on top of a balance. The beaker is on a heatproof mat. The water is then brought to boiling point with a heater. At the moment the water boils, the joulemeter is reset to zero.
● The heater is then allowed to boil the water for 5 minutes.
● The following measurements are taken:
  i)   joulemeter reading after 5 minutes – 60 kJ
  ii) mass of beaker and contents at the start – 968 g
  iii) mass of beaker and contents after 5 minutes – 944 g.

Calculation

$$\text{Mass of water turned to steam} = 968\,\text{g} - 944\,\text{g}$$
$$= 24\,\text{g} = 0.024\,\text{kg}$$
$$E = m\,L$$
$$\text{So } 60\,000 = 0.024 \times L$$
$$L = \frac{60\,000}{0.024}$$
$$= 2\,500\,000\,\text{J/kg}$$
$$= 2.5\,\text{MJ/kg}$$

▲ **Figure 3.13** The cooling curve for water.

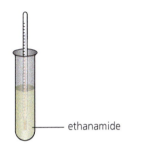

ethanamide

▲ **Figure 3.14**

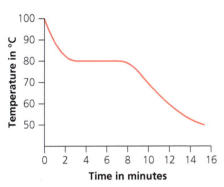

▲ **Figure 3.15** The cooling curve for ethanamide.

▲ **Figure 3.16** We cool when water evaporates from our bodies.

▲ **Figure 3.17**

Cooling graphs

When a boiling tube containing water is heated and then left to cool down, the temperature of the water drops gradually (see Figure 3.13).

When the temperature of the water is high, the temperature drops quickly. When water is closer to room temperature, the temperature drops more slowly.

If a substance changes state as it cools, the cooling curve takes a different shape. Ethanamide is a substance that melts at 80 °C. When a boiling tube containing ethanamide is allowed to cool from 100 °C, it cools quickly from 100 °C to 80 °C. Then the temperature remains constant for a few minutes as the ethanamide solidifies (or freezes). Although the boiling tube continues to transfer energy to the surroundings, the temperature of the ethanamide remains constant at 80 °C. This is possible because the ethanamide releases energy as the internal energy of its molecules decreases. When all the ethanamide has solidified, its temperature begins to fall again.

Test yourself

13 a) When we take exercise we sweat. The sweat evaporates from our skin. Why does sweating help us stay cool?

b) Even on a warm day having wet skin can soon make you feel cold. Explain why.

14 A solid is heated at a constant rate until it becomes a gas. Figure 3.18 shows how the temperature of the substance increases with time.

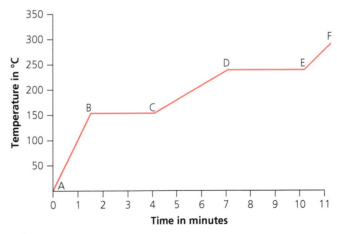

▲ **Figure 3.18**

a) Explain why the temperature of the substance remains constant over the periods:

i) B to C

ii) D to E.

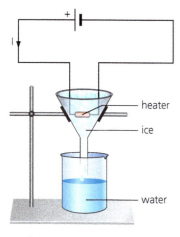

▲ Figure 3.19

b) At what temperature does the substance melt?

c) Which specific latent heat is higher, fusion or vaporisation? Give a reason for your answer.

15 A student designs an experiment to calculate the specific latent heat of fusion of ice. He uses a heater to melt ice as shown in Figure 3.19.

a) Explain why the ice should be allowed to reach 0 °C rather than being taken straight out of the freezer at a temperature of –18 °C.
The heater is turned on for 1 minute and 8 g of ice melts. The heater has a power of 50 W.

b) Calculate the energy supplied by the heater in 1 minute.

c) Calculate the specific latent heat of fusion of ice. Give your answer in joules per kilogram.

d) Give two reasons why the value obtained from this experiment is likely to be inaccurate.

Show you can...

Show you understand about latent heat by completing this task. Explain to a friend why it is much more painful to be burnt by 1 gram of steam at 100 °C than to be burnt by 1 gram of water at 100 °C.

Particle model and pressure

◯ The particle model of gases

As a result of studying the behaviour of gases, we have built up a model (or theory) to help us understand, explain and predict the properties of gases. This is called the **particle model** or **kinetic theory** of gases. The main points of the model are listed here.

- The particles in a gas (molecules or atoms) are in a constant state of random motion.
- The particles in a gas collide with each other and the walls of their container without losing any of their kinetic energy.
  - The temperature of the gas is related to the average kinetic energy of the molecules.
  - As the kinetic energy of the molecules increases the temperature of the gas increases.

Gas pressure

When the particles of a gas collide with a wall of their container, the particles exert a force on the wall. Figure 3.20 shows three particles bouncing off a container wall. Each particle exerts a force on the wall at right angles to the wall.

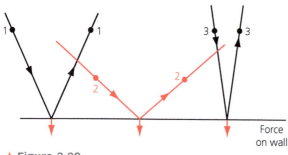

▲ Figure 3.20

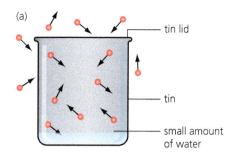

(a)

tin lid

tin

small amount of water

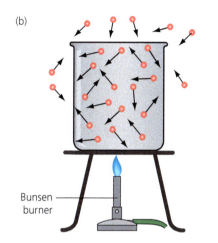

(b)

Bunsen burner

▲ **Figure 3.21** Safety note: This experiment should be done behind a safety screen and everyone should wear safety glasses.

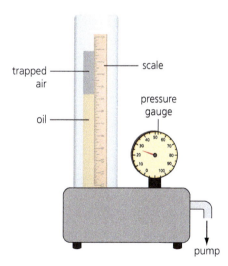

trapped air

oil

scale

pressure gauge

▲ **Figure 3.22**

pump

The pressure inside a container of gas, with a fixed volume, is increased when the temperature of the gas is increased. When the temperature of a gas increases, the average speed of the particles in the gas increases. This means that the particles hit the walls of the container with a greater force and the particles hit the walls more frequently. So the pressure increases.

Demonstrating gas pressure

Your teacher might demonstrate the effect of gas pressure as follows:

- Take a tin with a close fitting lid and put a small amount of water in it. Press the lid firmly in place.
- Then put the tin on a tripod and heat with a Bunsen burner. After a while the lid flies off.

So why does the lid fly off?

- In Figure 3.21a) the molecules inside the tin move at the same speed as the molecules outside the tin. There is no resultant force on the tin lid.
- In Figure 3.21b) two things have happened. As the tin warms up, some water evaporates so the number of molecules of gas inside the tin increases. Then the molecules travel faster as the temperature rises (shown with longer arrows on the molecules). The molecules inside the tin exert a force on the lid large enough to blow it off.

Expanding and compressing gases

The pressure in a gas can also be changed without changing the temperature. Figure 3.22 shows apparatus that can be used to change the pressure of a fixed mass of air at constant temperature.

By increasing the pressure on the oil using a pump, the column of trapped air is compressed. Figure 3.23 shows the relationship between the pressure of the trapped air and the length of the column of air.

The relationship between the volume of a fixed mass of gas, held at a constant temperature, and the pressure of the gas is shown by this equation:

$$P\,V = \text{constant}$$

pressure × volume = constant

Where pressure is in pascals, Pa
volume is in metres cubed, m^3.

This means that the pressure of the gas and the volume of the gas are inversely proportional.

$$P \propto \frac{1}{V}$$

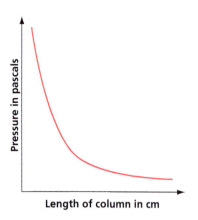

▲ Figure 3.23

Example

The gas inside a container of volume 4×10^{-3} m^3 is 2×10^5 Pa. The top of the container is sealed by a moveable piston. Pushing the piston inwards reduces the volume of the gas to 1×10^{-3} m^3. Calculate the pressure now exerted by the gas. The temperature does not change.

Answer

$P\,V = $ constant

So $P_1\,V_1 = P_2\,V_2$

$(2 \times 10^5) \times (4 \times 10^{-3}) = P_2 \times (1 \times 10^{-3})$

$8 \times 10^2 = P_2 \times 10^{-3}$

$P_2 = 8 \times 10^5$ Pa $= 800$ kPa

So for example:

- halving the volume of the gas causes the pressure to double
- doubling the volume of the gas causes the pressure to halve.

Explaining pressure in terms of gas particles

A fixed mass of gas means that the number of particles in the gas does not change. If the gas has a constant temperature, then the average kinetic energy (and speed) of the particles is also constant.

If the volume of a gas is reduced, the same number of particles hit a smaller area of container walls. The particles hit the walls of the container at the same speed but more frequently. This means that the force exerted on the walls increases. So if the force increases and the area is smaller, the pressure (which is the force per unit area) increases.

The opposite is true if the volume of a gas is increased. Now the particles hit a larger area of container walls. The particles still hit the walls at the same speed but this time less frequently. So now the force exerted on the walls decreases. This means that a smaller force acts on a larger area causing the pressure to decrease.

Test yourself

16 a) Describe the motion of the molecules in a gas.
   b) What happens to the motion of the molecules when the temperature of the gas is increased?
17 A cylinder of gas that is caught in a burning building can be a significant hazard. The temperature in a fire can rise as high as 1200 °C.
   a) Explain, in terms of molecular motion, why the pressure rises if the cylinder is in a fire.
   b) Explain why the cylinder is a danger to any fire fighters.
18 Use the idea of molecular motion to explain why:
   a) a gas exerts a pressure on the walls of its container
   b) when the temperature of a fixed volume of gas increases, the pressure increases.
19 Figure 3.24 shows a gas holder. It is filled with gas from the national pipeline where the pressure is 800 kPa. In the gas holder the volume of the gas is 240 000 m^3 and the pressure is 100 kPa. Calculate the volume of the gas when it was in the pipeline.

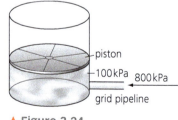

▲ Figure 3.24

Show you can...

Show that you understand the particle model of gases by completing this task. When a cylinder of gas is compressed into a smaller volume, the pressure exerted by the gas on the cylinder wall increases. Explain why.

○ Work and energy

When you use a bicycle pump to pump up your tyres, you find that the end of the pump gets hot. You can understand this in terms of molecular motion. As the piston moves forwards, inside the pump, it collides with molecules. The molecules then bounce off the piston at a faster speed. The kinetic energy of the molecules is now greater, and the temperature rises. You only notice a temperature rise in the air if you move the pump quickly. If you pump the tyres slowly, the air in the pump has time to cool down.

We can also explain the temperature rise in the air using the idea of **work**. Work is the transfer of energy by a force.

work done = force × distance moved in the direction of the force.

- When you push the pump, you do work.
- Doing work on a gas increases the internal energy of the gas.
- As the internal energy of the gas increases, the temperature of the gas rises.

When we drive a car, we are getting cylinders of gas to do work for us. In a petrol engine cylinder, a mixture of air and vaporised fuel is ignited by a spark. As the gas explodes the temperature rises and the pressure of the gas increases. The pressure in the gas exerts a force on the piston to push it outwards. The piston does work to push the car forwards.

We can explain the energy transfers involved in driving a car as follows:

The chemical energy store of the fuel decreases. Energy is transferred by mechanical work into the kinetic energy store of the car. Energy is also transferred by heating to the thermal energy store of the surroundings.

Test yourself

20 a) When a gas is rapidly compressed by a piston, the temperature of the gas rises. Give a reason why.
   b) When a gas expands rapidly the temperature of the gas falls. Give a reason why.

Chapter review questions

1 The sides of a block of wood measure 4.0 cm, 3.0 cm and 5.0 cm. The block of wood has a mass of 30.0 grams.

   a) Calculate the volume of the wood in m³.

   b) Calculate the density of the wood in kg/m³.

2 A man buys a 'gold' ornament from an antiques shop. He decides to check if the ornament is made of solid gold.

   The results of his measurements are shown below.
- mass of ornament = 320 g
- volume of ornament = 26×10^{-6} m³

   a) Explain how the man might have measured the mass and volume of the ornament.

   b) Calculate the density of the ornament.

   c) Use the data below to suggest what the man might find if he cuts his ornament in half.
- density of gold = 19 300 kg/m³
- density of lead = 11 600 kg/m³.

3 When a drop of ether is placed on the skin, the skin feels cold. Explain why.

4 a) State one difference between the arrangement of the molecules in water and the molecules in ice. Draw a diagram to illustrate your answer.

   b) An ice cube at a temperature of 0 °C is more effective in cooling a drink that the same mass of water at 0 °C. Give the reason why.

   c) Give the reason why ice floats on water.

5 Use the information in the table on page 74 to help you answer these questions.

   a) A heater supplies 200 kJ of energy to 40 kg of dry air at a temperature of 15 °C. Calculate the temperature of the air after it has been heated.

   b) A block of concrete has a mass of 60 kg. It cools down from 48 °C to 13 °C. Calculate the energy transferred from the block.

6 a) Explain in terms of molecular motion why a gas exerts a pressure on the walls of its container.

   b) The pressure in a container of gas increases when the temperature increases. Explain why.

7 A pan of water is placed on top of a cooker. The cooker transfers energy to the water at a rate of 500 W.

   a) When the water is boiling, the pan is left on the cooker for 5 minutes. Calculate the energy transferred to the cooker in this time.

   b) Calculate the mass of water that turns into steam in 5 minutes. (The specific latent heat of vaporisation of water is 2.5 MJ/kg.)

8 A cylinder of gas has a volume of 0.8 m³ at a pressure of 600 kPa. Calculate the volume occupied by the gas when the valve on the cylinder is opened and the gas is allowed to escape and reach atmospheric pressure of 100 kPa.

Practice questions

1 Which of the following is the correct unit for density?

kg/m² kg/m kg/m³ m³/kg [1 mark]

2 Figure 3.25 represents four measuring cylinders each containing a liquid. The mass and volume of the liquid in each cylinder are given.

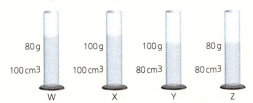

80 g 100 g 100 g 80 g

100 cm³ 100 cm³ 80 cm³ 80 cm³

W X Y Z

▲ Figure 3.25

Which two measuring cylinders could contain the same liquid?

a) W and X

b) W and Y

c) X and Y

d) X and Z [1 mark]

3 Each of the following statements describes either a solid, a liquid or a gas.

Copy each statement and write the correct words, solid, liquid or gas, next to each one.

a) It takes the shape of its container, but does not always fill the container. [1 mark]

b) The particles are in a regular pattern. [1 mark]

c) The particles are free to move over each other. [1 mark]

d) The particles move in random directions. [1 mark]

e) The particles always fill the whole of their container. [1 mark]

f) The particles vibrate about a fixed position. [1mark]

4 Figure 3.26 shows a bicycle pump. A student pushes his thumb tightly against the end of the pump, so no air can escape.

What will happen to each of the following when the student pushes the piston of the bicycle pump in?

(Assume the temperature of the air inside the pump does not change.)

a) The number of particles inside the pump. [1 mark]

b) The volume of air in the pump. [1 mark]

c) The pressure of the air in the pump. [1 mark]

d) The density of the air in the pump. [1 mark]

5 Energy is supplied to a substance and its temperature remains the same. Explain how this is possible. [2 marks]

6 Describe the differences between the arrangement of the atoms in a solid and in a gas. [4 marks]

7 Explain how you would use a measuring cylinder, electronic balance and some glass marbles to calculate the density of glass. [4 marks]

8 The apparatus in Figure 3.27 is used to heat up a block of metal of mass 2 kg. When the heater is turned on, the temperature of the block of metal increases as shown in Figure 3.28.

a) Use the graph to determine the temperature rise of the metal in the first 10 minutes of heating. [1 mark]

b) During 10 minutes of heating, 48 000 J of energy is supplied to the block. Calculate the specific heat capacity of the block. [4 marks]

c) Use the information in part (b) and information from the graph to show that the power of the heater is 80 W. [3 marks]

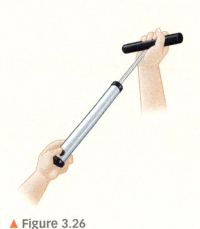

▲ Figure 3.26

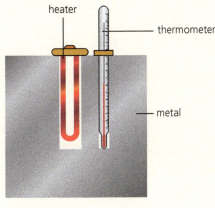

heater

thermometer

metal

▲ Figure 3.27

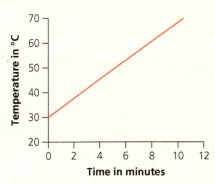

▲ Figure 3.28

9 Figure 3.29 shows a heater at the bottom of a boiling tube of solid wax. The heater is then connected to a power supply. A joulemeter measures the energy supplied to the heater as it melts the wax. The graph in Figure 3.30 shows how the temperature of the wax changes with the energy supplied.

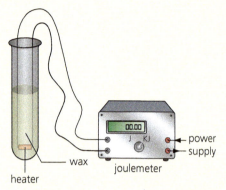

▲ Figure 3.29

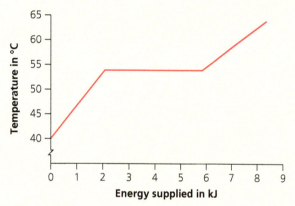

▲ Figure 3.30

a) State the melting temperature of wax. [1 mark]

b) The temperature of the wax remains constant as the wax melts. Explain why. [2 marks]

c) The mass of the wax is 50 g. Use information from the graph to calculate the specific latent heat of fusion for the wax. Give the unit. [4 marks]

10 This question is about a scuba diver and the pressure in his body. You need this information to help you answer the questions that follow.

The density of water is 1000 kg/m³.

The pressure of the air on the day of the dive was 100 kPa.

a) i) The diver descends to a depth of 25 m. Calculate the extra pressure on him due to the water. [3 marks]

ii) Show that the total pressure on the diver at a depth of 25 m is 345 kPa. [1 mark]

b) When the diver is at a depth of 25 m the air in his lungs has a volume of 5 litres.

i) Calculate the volume this air would occupy at the atmospheric pressure of 100 kPa. [3 marks]

ii) Explain why divers must breathe out as they rise to the surface. [1 mark]

11 A balloon seller has a cylinder of helium gas which he uses to blow up balloons.

The volume of the cylinder is 0.10 m³.

The pressure of the helium gas inside the cylinder is 1.0×10^7 Pa.

The balloon seller fills each balloon to a volume of 1.0×10^{-2} m³ and a pressure of 1.2×10^5 Pa.

a) Explain, in terms of particles, how the helium in the cylinder exerts a pressure on the sides of the cylinder. [2 marks]

b) Work out the total volume that the helium gas will occupy at a pressure of 1.2×10^5 Pa. You can assume that the helium does not change temperature. [3 marks]

c) Work out the number of balloons of volume 1.0×10^{-2} m³ that the balloon seller can sell using one cylinder of helium. [2 marks]

Working scientifically

Physical models

We all know what a scale model is and you have probably seen a model aircraft. It may not have been an exact replica but you would be able to recognise it as an aircraft. If you can't, then the model probably needs replacing.

In science we often use a physical model to help visualise objects or systems that are too big, too small or impossible to see. A scale model of the Solar System helps us to understand the position and distances between the planets and the Sun.

In this chapter you have read about the particle model of a gas. Atoms and molecules are too small for us to see directly so a physical model can help us to visualise what is happening and understand the real thing.

1 Describe another physical model given in this chapter.

Scientific models

A scientific model is an idea or related group of ideas used to explain something in the real world. A model is sometimes called a theory. Some people think that the word 'theory' means a guess; it does not. A theory or model is the idea used to explain observations and patterns in data.

A scientific model should be able to:

▶ explain observations

▶ be used to predict outcomes

▶ fit with other ideas in science.

Scientists use graphs, diagrams, equations, computer graphics and physical structures to represent and communicate models to others.

▲ Figure 3.31 This model aircraft looks and flies just like the real thing.

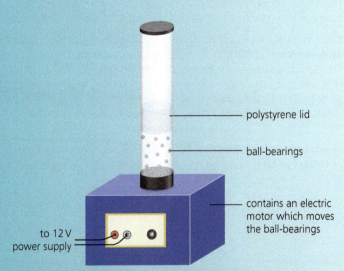

▲ Figure 3.32 The ball-bearings represent the atoms or molecules of a gas. The ball-bearings whizz around hitting the sides of the container. Turning the motor up is like increasing the temperature. The ball-bearings move faster and push the polystyrene lid upwards. So the volume expands but the pressure remains the same, just like a real gas.

polystyrene lid

ball-bearings

contains an electric motor which moves the ball-bearings

to 12 V power supply

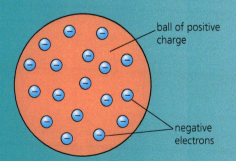

ball of positive charge

negative electrons

▲ Figure 3.33 A diagram used to represent the plum-pudding model of the atom (see page 91)

The particle model of a gas can be represented by mathematical equations. The equations can be used to explain and predict how changing one variable, for example the volume, affects the pressure and temperature of the gas.

2 The equation $F = ma$ (see page 159) is a mathematical model. What would you predict happens to the acceleration of an object when the resultant force on the object is increased? How would you be able to test this model?

Why do models change?

A scientific model is only as good as the evidence that supports it. A model is supported if any predictions made using the model turn out to be correct. However, a test that gives data which the model cannot predict provides evidence that the model may be wrong.

So models change to give a better fit to any new experimental results or observations.

In Chapter 4 you can read how new experimental evidence led to the plum-pudding model of the atom being replaced by the nuclear model. This model itself has been changed several times, each time to explain the most up-to-date observations.

3 What examples of changing models are given in Chapter 8?

Alternative models

Sometimes scientists have more than one model to explain the same thing. In Chapter 6 light is described as a wave. This is one model. However, sometimes light is better explained using the idea that it is a stream of particles. So is one model right and the other wrong? The answer is no. Each of the models is appropriate but in different situations.

Scientists know that they cannot explain everything. Sometimes there are alternative models but insufficient evidence to support or reject any of them. For example, how will the Universe end? Will it expand for ever, will it eventually shrink or will it stop expanding and stay at that size? A group of scientists using a new mathematical model believe the Universe will rip itself apart. However, there is no need to panic. The model predicts it will not happen for another 22 billion years.

4 What would be needed for one of the models explaining the end of the Universe to be accepted by scientists and the other models rejected?

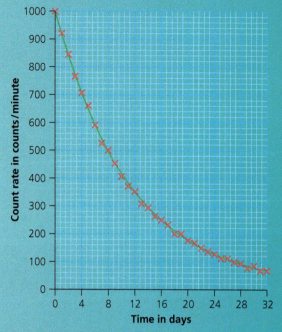

▲ **Figure 3.34** A graphical representation of the model used to explain the decay of a radioactive substance (see pages 99–100).

4 Atomic structure

Henry Becquerel discovered radioactivity in 1896. Over the last century, we have learnt how to use radioactive materials safely and to put them to good use. Radioactive sources are used in industry, agriculture and in medicine. The image shows the concentration of radioactive sugar 2 hours after tracer molecules were fed into a plant. The red colour shows a high concentration of sugar in the young leaves of the plant, which are growing. The fast growing young leaves take the sugar from the older leaves, which appear blue. Doctors use a similar method to look at fast-growing cancer tumours.

Specification coverage

This chapter covers the specification points: 4.4 Atomic structure. It covers 4.4.1 Atoms and isotopes, 4.4.2 Atoms and nuclear radiation, 4.4.3 Hazards and uses of radioactive emissions and of background radiation and 4.4.4 Nuclear fission and fusion.

Prior knowledge

Previously you could have learned

› All materials are made up of tiny particles called atoms.
› Elements are made up of only one type of atom.
› An atom has a very small positively charged nucleus.
› The nucleus contains protons and neutrons.
› Negatively charged electrons orbit the nucleus.
› The proton carries a positive charge and the electron carries a negative charge of the same size as the proton.
› An atom is neutral in charge, because the positive charge on the nucleus is balanced by the negative charge of the electrons.

Test yourself on prior knowledge

1 Which two particles are in the nucleus of an atom?
2 Which particle is more massive, the proton or the electron?
3 Why are the number of protons and electrons the same for a particular atom?

Atoms and isotopes

(a) hydrogen atom

(b) lithium atom

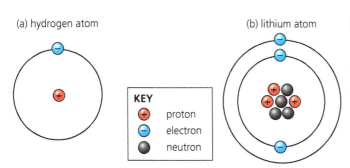

KEY
⊕ proton
⊖ electron
● neutron

▲ **Figure 4.1** These figures show the numbers of protons, neutrons and electrons in a hydrogen and lithium atom.

KEY TERMS

A **proton** is a positively charged particle which is found in the nucleus of an atom.

A **neutron** is a neutral particle found in the nucleus of an atom.

An **electron** is a negatively charged particle that orbits the nucleus of an atom.

○ Neutrons, protons and electrons

We cannot see atoms directly because they are so small. However, indirect measurements show us that the radius of an atom is about 10^{-10} m (0.000 000 000 1 m). The nucleus of the atom is much smaller than the atom. The radius of the nucleus is less than 1/10 000 of the radius of the atom.

Inside the nucleus there are two types of particle, **protons** and **neutrons**. The protons and neutrons have approximately the same mass. A proton has a positive charge and the neutron has no charge. Outside the nucleus there are **electrons** which orbit the nucleus at distances of about 10^{-10} m. Electrons are able to move in different orbits around the nucleus, and change their orbit by absorbing or emitting electromagnetic radiation.

Electrons have very little mass in comparison with protons and neutrons. Electrons carry a negative charge which is the same size as the positive charge on the proton. Because protons and neutrons are much more massive than electrons, most of the mass of an atom is in its nucleus.

Example

A gold atom has a radius of 1.34×10^{-10} m and a gold nucleus has a radius of 7.3×10^{-15} m. How many times larger is the radius of a gold atom than the radius of a gold nucleus?

Answer

$$\frac{\text{radius of gold atom}}{\text{radius of gold nucleus}} = \frac{1.4 \times 10^{-10}\,\text{m}}{7 \times 10^{-15}\,\text{m}}$$

$$= 20\,000$$

This answer gives an approximate ratio of the atomic and nuclear radii for gold. This ratio is different for different elements.

○ Atoms and ions

A hydrogen atom has one proton and one electron; it is electrically neutral because the charges of the electron and proton cancel each other out. A helium atom has two protons and two neutrons in its nucleus, and two electrons outside that. The helium atom is also neutral because it has the same number of electrons as it has protons.

Ions

Atoms are electrically neutral because the number of protons balances exactly the number of electrons. However, it is possible either to add extra electrons to an atom or to take them away. When an electron is added to an atom a **negative ion** is formed; when an electron is removed a **positive ion** is formed. Ions are made in pairs because an electron that is removed from one atom attracts itself to another atom, so a positive and negative ion pair is formed.

○ Atomic and mass numbers

The number of protons in the nucleus of an atom determines what element it is. Hydrogen atoms have one proton, helium atoms two protons, uranium atoms 92 protons. The number of protons in the nucleus decides the number of electrons surrounding the nucleus. The number of protons in the nucleus is called the **atomic number** of the atom (symbol Z). So the proton number of hydrogen is 1; Z = 1.

The mass of an atom is determined by the number of neutrons and protons added together. Scientists call this number the **mass number** of an atom.

atomic number = number of protons

mass number = number of protons plus neutrons

For example, an atom of sodium has eleven protons and twelve neutrons. So its atomic number is 11 and its mass number is 23. To save time in describing we can write it as $^{23}_{11}$Na; the mass number appears on the left and above the symbol Na, for sodium, and the atomic number on the left and below.

TIP

The **radius of a nucleus** is less than 1/10 000 of the **radius of an atom**.

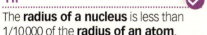

KEY TERM

atomic number = number of protons
mass number = number of neutrons plus protons

Mass number $^{23}_{11}$Na
Atomic number

KEY TERM

Isotopes are different forms of a particular element. Isotopes have the same number of protons but different numbers of neutrons.

TIP

The symbols $^{23}_{11}$Na describe only the nucleus of a sodium atom.

○ Isotopes

Not all the atoms of a particular element have the same mass. For example, two sodium atoms might have mass numbers of 23 and 24. The nucleus of each atom has the same number of protons, 11, but one atom has 12 neutrons and the other 13 neutrons. Atoms of the same element (sodium in this case) that have different masses are called **isotopes**. These two isotopes of sodium can be written as sodium-23, $^{23}_{11}$Na, and sodium-24, $^{24}_{11}$Na.

Test yourself

1 a) What is the approximate radius of an atom?

| 10^{-3} m | 10^{-6} m | 10^{-10} m |

  b) Use an answer from the box to complete the sentence.

| 1000 | 10 000 | 100 000 |

  The radius of an atom is about _____ times larger than the radius of a nucleus.

2 The diagram shows the nuclei of three atoms. Which two atoms are isotopes of the same element? Give a reason for your answer.

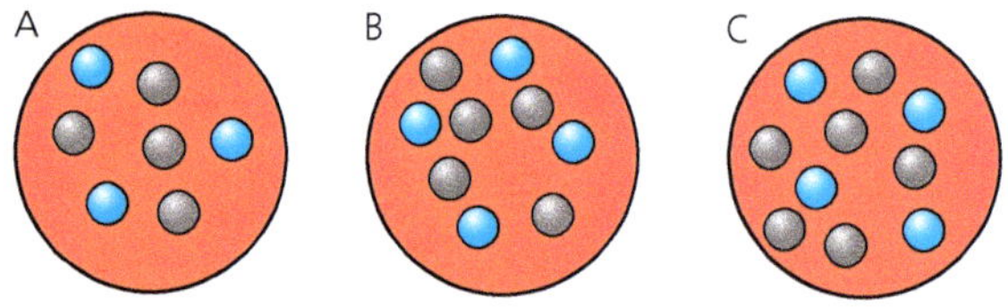

▲ Figure 4.2

3 a) A nitrogen atom has 7 protons, 7 neutrons and 7 electrons.

   i) What is the atomic number of nitrogen?

   7 14 21

   Give a reason for your answer.

   ii) What is the mass number of nitrogen?

   7 14 21

   Give a reason for your answer.

  b) Explain why a nitrogen atom is neutral.

4 Calculate the number of protons and neutrons in each of the following nuclei.

  a) $^{17}_{8}$O   b) $^{200}_{80}$Hg   c) $^{238}_{92}$U   d) $^{3}_{1}$H

5 Gadolinium-156 and gadolinium-158 are two isotopes of gadolinium, which has an atomic number of 64.

  a) Explain what the numbers 64, 156 and 158 mean.

  b) i) What do these two isotopes have in common?

     ii) How are the two isotopes different?

6 Explain the term isotope.

7 The radius of a magnesium nucleus is 3.0×10^{-15} m and the radius of a magnesium atom is 1.5×10^{-10} m. How many times larger is the radius of a magnesium atom than the radius of a magnesium nucleus?

Show you can...

Complete this task to show that you understand the model of the atom.

Write a paragraph to explain and describe the structure of an atom. In your answer include these words: nucleus, proton, neutron, electron, mass number and atomic number.

◯ Discovery of the nucleus

The discovery of the nucleus and the development of our understanding of the atom provide a clear example of how new experimental evidence can lead to a scientific model being changed or replaced.

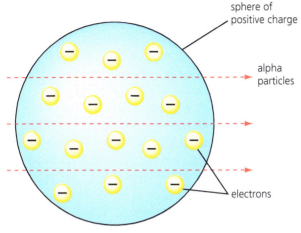

▲ **Figure 4.3** The 'plum pudding' atomic model.

Democritus, a Greek philosopher, lived from 460–370 BC. He was the first person to suggest the idea of atoms as small particles that cannot be cut or divided.

The discovery of the electron

In 1897 J J Thompson discovered that electrons were emitted from the surface of hot metals. Thompson showed that electrons are negatively charged and that they are much less massive than atoms. This discovery led to a change in the accepted atomic theory.

In 1904 Thompson proposed a new model for the atom. His idea was that atoms were made up of a ball of positive charge with electrons dotted around inside it (Figure 4.3). This idea is known as the 'plum pudding' model of the atom as it looks rather like a solid pudding with plums in it.

The nuclear model of the atom

In 1909 Geiger and Marsden discovered a way of exploring the insides of atoms. They directed a beam of alpha particles at a thin sheet of gold foil. Alpha particles were known to be positively charged helium ions, He^{2+}, which travel very quickly. They had expected all of these energetic particles to pass straight through the thin foil because they thought the atom was like Professor Thompson's soft plum pudding model. Much to their surprise they discovered that although most of them travelled through the foil without any noticeable change of direction, a very small number of the alpha particles bounced back.

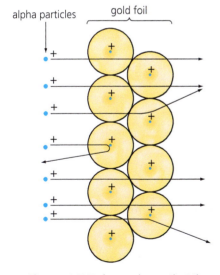

▲ **Figure 4.5** Today we know that the nucleus is positively charged. Alpha particles which are deflected by a large amount are repelled by the strong electric field of the nucleus.

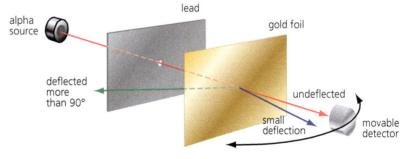

▲ **Figure 4.4** Geiger and Marsden's experiment.

In 1911, Ernest Rutherford produced a theory to explain these results. This is illustrated in Figure 4.5.

Rutherford suggested that the deflection of an alpha particle was due to an electrostatic interaction between it and a very small charged nucleus. He also suggested that the nucleus must be massive, because it did seem to be moved by the energetic alpha particle. In Rutherford's original paper he suggested that the nucleus might be either positively or negatively charged. A positive nucleus would repel the positively charged alpha particle backwards, and a negative nucleus might pull the alpha particle round it, in the same way that a comet falling towards the Sun has its direction changed by the pull of gravity.

Alpha scattering explained

Now we know that the gap between the nucleus and electrons is large; the diameter of the atom is about 20 000 times larger than the diameter of the nucleus itself (Figure 4.6). After Rutherford's original paper, scientists confirmed that the nucleus is positively charged, and that allows us to explain the scattering of the particles as follows. Because so much of the atom is empty space (Figure 4.6), most of the alpha particles could pass through it without getting close to the nucleus. Some particles pass close to the nucleus and so the positive charges of the alpha particle and the nucleus repel each other, causing a small deflection. A small number of particles met the nucleus head on; these are turned back the way they came (Figure 4.5). The fact that only a very tiny fraction of the alpha particles bounce backwards tells us that the nucleus is very small indeed. Rutherford proposed that all the mass and positive charge of an atom are contained in the nucleus and that the electrons outside the nucleus balance the charge of the protons.

Later experiments led to the idea that the positive charge of any nucleus could be subdivided into a whole number of smaller particles. These particles were called protons, and the positive charge on a proton was discovered to be the same size as the negative charge on an electron. Rutherford did not know that there are neutrons in the nucleus; these were discovered by James Chadwick in 1932. Chadwick's discovery allowed scientists to account for the mass of the atom.

The Bohr model of the atom

In 1913 Niels Bohr suggested a model of the atom in which electrons move round the nucleus in circular orbits. In this model electrons can change their orbits. Figure 4.7 shows the Bohr model of a hydrogen atom, with the first three energy levels.

The Bohr model was successful in explaining why hydrogen emits particular wavelengths of electromagnetic radiation. In Figure 4.7 an electron emits electromagnetic radiation and so loses some energy. It moves closer to the nucleus as it falls from level 3 to level 2. The reverse process is possible too: if an electron absorbs electromagnetic radiation it can move into a higher energy level further away from the nucleus.

Although the Bohr model was successful up to a point, it does not allow a full explanation of the behaviour of electrons in larger atoms. However, Bohr adapted his model for the hydrogen atom by suggesting that electrons in larger atoms are also confined to specific orbits. The most recent theories suggest that electrons move in very complex orbits, which have a variety of different shapes. However, once an electron is in an orbit, it has a fixed amount of energy.

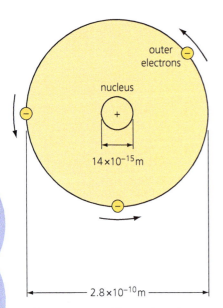

▲ **Figure 4.6** A gold atom (not drawn to scale). The diameter of the atom is about 20 000 times larger than the diameter of the nucleus (though this ratio is not the same for all elements).

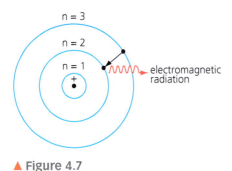

▲ **Figure 4.7**

TIP ✓

Alpha particles are repelled or deflected by the nucleus of an atom because they both have a positive charge.

TIP ✓

Rutherford's theory states that the nucleus is:
• positively charged
• very small
• contains nearly all the mass of the atom.

The theory suggests that electrons orbit around the nucleus at distances much greater than the radius of the nucleus.

Test yourself

8 a) How is the mass of the atom distributed in the plum pudding model?
   b) Where is most of the mass of the atom in the nuclear model?
9 Explain how you know that the polarity of the charge on an alpha particle is the same as that of the nucleus.
10 Why did most of the alpha particles pass through the foil without being deflected?
11 Describe the plum pudding model of the atom.
12 a) Describe Bohr's model for the atom. Draw a diagram to help your explanation.
   b) Explain what happens to an electron when
      i) it absorbs electromagnetic radiation.
      ii) it emits electromagnetic radiation.
13 Figure 4.8 shows the path of an alpha particle being deflected by the charge of a gold nucleus.

   Make a copy of the diagram to show the paths of two more alpha particles B and C.

Show you can...

Complete this task to show you understand how new evidence may lead to a scientific model being changed or replaced.

In the nineteenth century, scientists thought that atoms were the smallest particles. Explain what evidence led scientists to change this model of the atom.

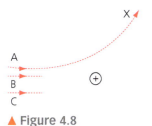

▲ Figure 4.8

Atoms and radiation

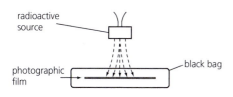

radioactive source

photographic film

black bag

▲ **Figure 4.9** How Becquerel discovered radioactivity. Radioactive particles pass through a lightproof bag to expose a photographic film.

Henri Becquerel discovered radioactivity in 1896. He placed some uranium salts next to a photographic plate which had been sealed in a thick black bag to prevent light exposing the plate. When the plate was later developed it had been affected as if it had been exposed to light. Becquerel realised that new particles were emitted from uranium salts, which passed through the bag.

○ Nuclear decay

The nuclei of most atoms are very stable. The atoms that we are made of have been around for thousands of millions of years. Atoms may lose or gain a few electrons during chemical reactions, but the nucleus does not change during such processes.

However, there are some atoms that have unstable nuclei which throw out particles to make the nucleus more stable. This is a random process that depends only on the nature of the nucleus. The rate at which particles are emitted from a nucleus is not affected by other factors such as temperature or chemical reactions. One element discovered that emits these particles is radium, and the name **radioactivity** is given to this process.

KEY TERMS

An **alpha particle** is formed from two protons and two neutrons.
A **beta particle** is a fast moving electron.
A **gamma ray** is an electromagnetic wave.

TIP

When a beta particle is emitted from a nucleus, a neutron turns into a proton. The mass number of the nucleus remains the same, but the atomic number increases by 1.

There are four types of radioactive emission.

Alpha particles are identical to the nuclei of helium atoms. The alpha particle is formed from two protons and two neutrons, so it has a mass number of four and an atomic number of two. When an alpha particle is emitted from a nucleus it causes the nucleus to change into another nucleus with a mass number four less and an atomic number two less than the original one, for example:

$$\overset{238}{\underset{92}{}}U \quad \rightarrow \quad \overset{234}{\underset{90}{}}Th \quad + \quad \overset{4}{\underset{2}{}}He$$

| uranium nucleus | thorium nucleus | alpha particle (helium nucleus) |

This is called **alpha decay**.

Beta particles are fast moving electrons. In a nucleus there are only protons and neutrons, but a beta particle is made and ejected from a nucleus when a neutron turns into a proton and an electron. Since an electron has a very small mass, when it leaves a nucleus it does not alter the mass number of that nucleus. However, the electron carries away a negative charge so the removal of an electron increases the atomic number of a nucleus by 1. For example, carbon-14 decays into nitrogen by emitting a beta particle:

$$\overset{14}{\underset{6}{}}C \quad \rightarrow \quad \overset{14}{\underset{7}{}}N \quad + \quad \overset{0}{\underset{-1}{}}e$$

| carbon nucleus | nitrogen nucleus | beta particle (electron) |

When some nuclei decay by sending out an alpha or beta particle, they also give out a **gamma ray**. Gamma rays are electromagnetic waves, like radiowaves or visible light. They carry away from the nucleus a lot of energy, so that the nucleus is left in a more stable state. Gamma rays have no mass or charge, so when one is emitted there is no change to the mass or atomic number of a nucleus.

Neutrons are emitted from some highly unstable nuclei. The effect of this is to reduce the mass number by 1, but the atomic number does not change. Neutron emission is rare, but neutrons are a dangerous radiation. You will meet them again when you learn about fission. An example of neutron emission from helium-5 is given below:

$$\overset{5}{\underset{2}{}}He \quad \rightarrow \quad \overset{4}{\underset{2}{}}He \quad + \quad \overset{1}{\underset{0}{}}n$$

A summary of the four types of radioactive emission.

	Radiation emitted from nucleus	Change in mass number	Change in atomic number
alpha (α) decay	Helium nucleus $\overset{4}{\underset{2}{}}He$	−4	−2
beta (β) decay	Electron $\overset{0}{\underset{-1}{}}e$	0	+1
gamma (γ) decay	Electromagnetic waves	0	0
neutron (n) decay	Neutron	−1	0

TIP

Make sure you know and understand the nature of alpha, beta, gamma and neutron radiations.

Ionisation

All types of radiation cause **ionisation**. This is why we must be careful when we handle radioactive materials. The radiation makes ions in our bodies and these ions can then damage our body tissues.

Your teacher can show the ionising effect of radium by holding some close to a charged gold leaf electroscope (Figure 4.11). The electroscope is charged positively at first so that the gold leaf is repelled from the metal stem. When a radium source is brought close to the electroscope, the leaf falls, showing that the electroscope has been discharged. The reason for this is that the alpha particles from the radium create ions in the air above the electroscope. This is because the charges on the alpha particles pull some electrons out of air molecules (Figure 4.10). Both negative and positive ions are made; the positive ones are repelled from the electroscope, but the negative ones are attracted so that the charge on the electroscope is neutralised. It is important that you understand that it is not the charge of the alpha particles that discharges the electroscope, but the ions that they produce.

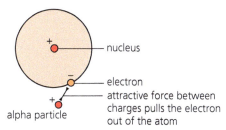

▲ **Figure 4.10** Alpha particles cause heavy ionisation.

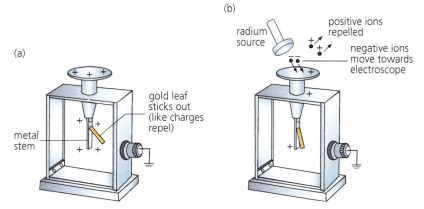

▲ **Figure 4.11** a) Positively charged electroscope. b) Negative ions neutralise the electroscope. **Safety note**: we always handle a radioactive source using tongs, to keep the source away from our body.

Test yourself

14 What is the nature of each of the following:
   a) an alpha particle
   b) a beta particle
   c) a gamma ray?
15 Which row in the table, A, B or C, shows what happens to the mass number and atomic number of a nucleus when a gamma ray is emitted from it?

	Mass number	Atomic number
A	Increases	Decreases
B	Does not change	Does not change
C	Decreases	Increases

Show you can...

Complete this task to show you understand the nature of radioactivity.

Explain the nature of alpha, beta, gamma and neutron radiation.

State examples of balanced nuclear equations for the emission of alpha and beta particles from a nucleus.

16 a) What is meant by the term ionisation?

b) Why are ions always produced in pairs?

17 Fill in the gaps in the following radioactive decay equations:

a) $^{3}_{?}\text{H} \rightarrow \, ^{?}_{2}\text{He} + \, ^{0}_{?}\text{e}$

b) $^{229}_{90}\text{Th} \rightarrow \, ^{?}_{?}\text{Ra} + \, ^{4}_{2}\text{He}$

c) $^{14}_{6}\text{C} \rightarrow \, ? + \, ^{0}_{-1}\text{e}$

d) $^{209}_{82}\text{Pb} \rightarrow \, ^{?}_{83}\text{Bi} + \, ?$

e) $^{225}_{89}\text{Ac} \rightarrow \, ^{?}_{87}\text{Fr} + \, ?$

18 $^{238}_{92}\text{U}$ decays by emitting an alpha particle then two beta particles. Which element is produced after these three decays?

19 Explain how a radioactive source emitting alpha particles can discharge a negatively charged electroscope.

○ Detecting particles

We make use of the ionising properties of alpha, beta and gamma radiations to detect them. Those three radiations are emitted by radioactive sources permitted in schools.

KEY TERMS

A **Geiger–Müller (GM)** tube is a device which detects ionising radiation. An electronic counter can record the number of particles entering the tube.

The **background count** is the average count rate which a Geiger–Müller tube records over a period of time when the counter is not close to a radioactive source. The background count is caused by radioactive materials in our environment.

Practical

The range and penetration of radiation

Safety note: Only your teacher can do this experiment as you are not allowed to handle radioactive sources until you are over 16.

Figure 4.12 shows how your teacher can use a **Geiger-Müller (GM)** tube and a radioactive source to investigate the range of different radiations.

1 The first step is to take account of the background radiation. This is the radiation from various sources that is around us all the time. By counting over a period of a minute, you can establish the **background count**.

2 A source emitting alpha particles, for example, can then be placed in front of the GM tube. By varying the separation (x) of the tube and the source, the range of the radiation may be calculated. When the count rate falls to the same as the background count, no radiation from the source is reaching the GM tube. This gives you the range of the radiation.

3 You can also check which materials stop a type of radiation. Now you keep the distance, x, constant. Then you can place various absorbers such as paper or metal foils in between the source and the GM tube.

The results of these experiments are summarised on page 97.

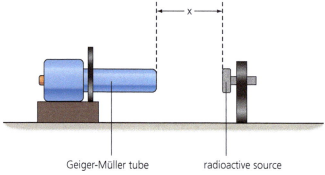

Geiger-Müller tube   radioactive source

▲ Figure 4.12

○ Properties of radiation

Alpha particles travel about 5 cm through air but can be stopped by a sheet of paper (Figure 4.13). They ionise air very strongly.

Beta particles can travel several metres through air. They are stopped by a sheet of aluminium that is a few millimetres thick (Figure 4.13). They do not ionise air as strongly as alpha particles.

Gamma rays can only effectively be stopped by a very thick piece of lead (Figure 4.13). Gamma rays only ionise air very weakly and travel great distances through air.

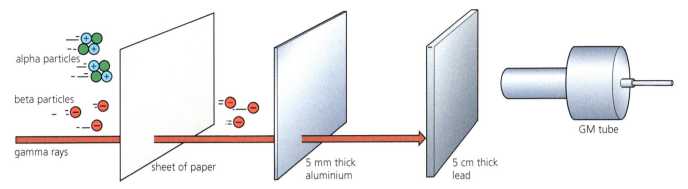

alpha particles

beta particles

gamma rays

sheet of paper

5 mm thick aluminium

5 cm thick lead

GM tube

▲ Figure 4.13

○ Radiation damage

If radiation gets into our body, damage can occur to cells and tissues. The ions which are produced by the radiation produce chemicals which destroy cells they come into contact with. Alpha particles cause the most damage if they get inside the body, as they are strongly ionising; this could happen if we breathed in a radioactive gas. An alpha source in school is less dangerous as the radiation only travels short distances, so does not enter our body.

Although gamma rays are less ionising than alpha particles, a gamma ray source in a laboratory is a hazard, as the rays are so penetrating. A gamma ray can pass into our body from a source several metres away from us.

The range and penetration of radiation.

Radiation	Nature	Range in air	Ionising power	Penetrating power
Alpha α	Helium nucleus	A few centimetres	Very strong	Stopped by paper
Beta β	Electron	A few metres	Medium	Stopped by aluminium
Gamma γ	Electromagnetic waves	Great distances	Weak	Stopped by thick lead

○ Background radiation

There are a lot of rocks in the Earth that contain radioactive uranium, thorium, radon and potassium, and so we are always exposed to some ionising particles. Radon is a gas that emits alpha particles. Since we can inhale this gas, it is dangerous, as radiation can get inside our lungs. In addition, the Sun emits lots of protons, which can also create ions in our atmosphere. These are two of the sources that make up background radiation. Figure 4.14 shows the contribution to the total background radiation from all natural sources in Britain. Fortunately, the level of background radiation is quite low and in most places it does not cause a serious health risk.

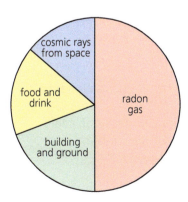

cosmic rays from space

food and drink

building and ground

radon gas

▲ Figure 4.14 Natural sources of background radiation in Britain.

KEY TERM ★

Background radiation comes from natural and man-made sources in our environment.

Test yourself

20 Which one of the following, A, B or C, is a property of beta radiation?
  A It is the most strongly ionising radiation.
  B It will travel through several metres of air.
  C It can be easily stopped by paper.

21 Which source of background radiation would give a person the largest dose of radiation?
cosmic rays   food and drink   radon gas

22 a) Which type of radiation has a range of a few centimetres in air?
  b) Which types of radiation are stopped by thin metal sheets?

23 Explain why a teacher uses long tongs to handle a source of radiation.

24 A radioactive source is placed 2 cm from a GM tube and a count rate of 120 counts per second is measured. When a piece of paper is put between the source and the GM tube, the count rate reduces to 75 counts per second. A 5 mm thick piece of aluminium between the source and tube reduces the measured count rate to the background level. Explain what this tells you about the source.

25 a) Explain what is meant by the term background radiation.
  b) Who is most likely to receive the highest annual radiation dose?
  an airline pilot   a motor mechanic   an office worker
  Give a reason for your answer.

26 a) Radon gas is a product of radioactive decay in granite rocks. Radon is also radioactive and emits alpha particles. Explain why radon might be a hazard to mining engineers working underground in granite rocks.
  b) Explain why a strong gamma source left in a laboratory might be dangerous to us.

Show you can...

Complete this task to show you understand how to investigate the properties of radiations.

Design an experiment to investigate the range of alpha particles in air. How would you adapt this experiment to investigate the penetrating power of alpha particles?

○ Radioactive decay

The atoms of some radioactive materials decay by emitting alpha, beta or gamma radiations from their nuclei. However, it is not possible to predict when the nucleus of one particular atom will decay. It could be the next second or sometime next week or not for a million years. Radioactive decay is a random process.

Random process

The radioactive decay of an atom is rather like rolling dice or tossing a coin. You cannot say with certainty that the next time you toss a coin it will fall heads up. However, if you throw a lot of coins you can start to predict how many of them will fall heads up. You can use this idea to help you understand how radioactive decay happens. Imagine you start off with a thousand coins. When any coin falls heads up then it has 'decayed' and you take it out of the game. The table on page 99 shows

▲ **Figure 4.15** On average how many sixes will turn up when you roll 100 dice? Why can you not be certain what will happen on each occasion?

the likely result (on average). Every time you throw a lot of coins, about half of them will turn up heads.

Radioactive decay

Radioactive materials decay in a similar way. If we start off with a million atoms then after a period of time (for example 1 hour), half of them will have decayed. In the next hour we find that half of the remaining atoms have decayed, leaving us with a quarter of the original number. The period of time taken for half the number of nuclei to decay in a radioactive sample is called the **half-life** and it is given the symbol $t_{\frac{1}{2}}$. It is important to understand that we have chosen a half-life here of 1 hour to explain the idea. Different radioisotopes have different half-lives.

> **KEY TERM**
>
> The **half-life** is the time taken for the number of nuclei in a radioactive isotope to halve. In one half-life the activity or count rate of a radioactive sample also halves.

Count rate and activity

The activity of a source is equal to the number of particles emitted per second. We can express this as counts per second, but in honour of Henry Becquerel, this unit is called the **becquerel** (Bq).

The count rate is the term we use when a GM tube is measuring the radiation emitted from a radioactive source. This is different from the activity of a source, because not all the radiation emitted from the source goes into the GM tube. Count rates may be in counts per second or sometimes counts per minute.

> **KEY TERM**
>
> 1 **becquerel** (1 Bq) = an emission of 1 particle/second.

Measurement of half-life

If you look at the table you can see that the number of nuclei that decayed in the first hour was 500 000, then in the next hour 250 000 and in the third hour 125 000. So as time passes not only does the number of nuclei left get smaller but so does the rate at which the nuclei decay. So by measuring the activity of a radioactive sample we can determine its half-life.

Modelling radioactive decay using coins.

Throw	Number of coins left
0	1000
1	500
2	250
3	125
4	62
5	31
6	16
7	8
8	4
9	2
10	1

Time (hour)	Number of nuclei left
0	1000000
1	500000
2	250000
3	125000
4	62500
5	31250
6	15620
7	7810
8	3900
9	1950
10	975

Figure 4.16 shows the result of a laboratory experiment to determine the half-life of a radioactive material. You can see that the count rate detected halves every half-life. At the start of the experiment the count

rate was 40 per second, after 50 seconds (one half-life) it has reduced to 20 per second, and after a further 50 seconds the count rate has halved again to 10 per second.

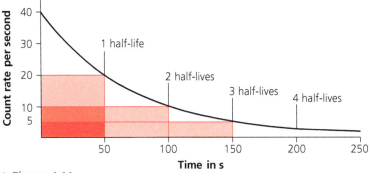

▲ Figure 4.16

Correcting for the background count

When we measure the activity of a weak radiation source we need to make a correction for background radiation before we calculate its half-life. In an experiment to find the half-life of a source we would take a background count before placing the source close to the GM tube. When we begin the experiment we need to remember that some of the count rate is due to background radiation. If the background count is 3 Bq and the activity measured by a counter is 43 Bq, then we know that only 40 Bq is due to the radioactive source.

▲ Figure 4.17

Practical

A model for radioactive decay

This practical task helps you to understand why the count rate from a radioactive source varies each time you try to count it.

Collect 100 dice. How many sixes would you expect to throw each time you roll the 100 dice? Roll the dice 10 times and record the number of sixes on each occasion.

Now explain why your teacher does not measure exactly the same count rate each time a GM tube is placed near a radioactive source.

Devise an experiment, using the dice, to model the decay of a radioactive source with a short half-life.

Carbon dating

Carbon-14, $^{14}_{6}C$, is a radioactive isotope; it decays to nitrogen with a half-life of about 5700 years. All living things (including you) have a lot of carbon in them, and a small fraction of this is carbon-14. When a tree dies, for example, the radioactive carbon decays and after 5700 years the fraction of carbon-14 in the dead tree will be half as much as you would find in a living tree. So by measuring the amount of carbon-14 in ancient relics, scientists can calculate their age.

TIP
Carbon dating is a useful way of dating same objects, but how the technique works is not featured in the AQA specification.

Half-life and stability

Radioactive isotopes have a wide range of half lives. Less stable isotopes decay more quickly, so they have shorter half-lives than more stable isotopes. Half-lives can be very long: for example uranium-238 decays with a half-life of 4.5 billion years. By contrast other half-lives are measured in fractions of a second. The table on page 101 lists the half-lives of some radioactive isotopes.

Half-lives of some radioisotopes.

Isotope	Half-life
Potassium-40	1.3 billion years
Carbon-14	5700 years
Caesium-137	30 years
Iodine-131	8 days
Lawrencium-260	3 minutes
Nobelium-252	2.3 seconds

Test yourself

27 Use a word from the box to complete the sentence.

count rate   half-life   reaction

The _____ is the number of alpha, beta or gamma emissions detected from a radioactive source in one second.

28 The graphs in Figure 4.18 show the decay of three different radioactive isotopes.
Which isotope has:
a) the longest half-life?
b) the shortest half-life?
c) the possibility of being used as a medical tracer?

29 Explain what the word random means.

30 A GM tube is placed in a laboratory to measure the background count. In five separate periods of 10 minutes the following counts were recorded: 147, 134, 129, 141, 119. Why are the counts different?

31 A radioactive material has a half-life of 15 minutes. What does this mean? How much of the original material will be left after 60 minutes?

32 The results in the table for the count rate of a radioactive source were recorded every minute.
a) A correction was made for background count. What does this mean?
b) Plot a graph of the count rate (*y*-axis) against time (*x*-axis) and use the graph to work out the half-life of the source.

33 A radioactive isotope has a half-life of 8 hours. At 12 noon on 2 March a GM tube measures a count rate of 2400 per second.
a) What will be the count rate at 4.00 am on 3 March?
b) At what time will a count rate of approximately 75 per second be measured?

34 a) Archaeologists are analysing ancient bones from a human settlement. They discover that a sample of bone has one-sixteenth of the carbon-14 of modern human bones. How old is the settlement? (The half-life of carbon-14 is 5700 years.)
b) Carbon dating is not accurate for ages greater than about 50 000 years. Explain why.

35 The age of rocks can be estimated by measuring the ratio of the isotopes potassium-40 and argon-40. We assume that when the rock was formed it was molten, and that any argon would have escaped. So, at the rock's formation there was no argon. The half-life of potassium-40 is 1.3×10^9 years and it decays to argon-40. Analysis of two rocks gives these potassium (K) to argon (A) ratios:

Rock X $\frac{K}{A} = \frac{1}{1}$; Rock Y $\frac{K}{A} = \frac{1}{7}$

Calculate the ages of the two rocks.

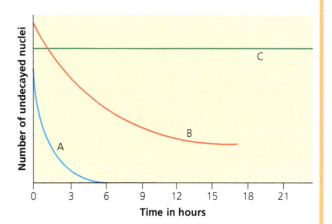

▲ Figure 4.18

Count rate in Bq	Time in minute
1000	0
590	1
340	2
200	3
120	4
70	5

Show you can...

Show you understand about radioactive decay by completing this task. A friend of yours has missed school. Write some brief notes to help him understand the meanings of these words: half-life; random; radiation.

The hazards of radiation

The Fukushima Daiichi power plant in Japan was devastated by the tsunami produced by the Tohoku earthquake of 11 March 2011. This led to the second biggest nuclear disaster in history, producing at its peak a local activity of 400 million million Bq. While nuclear power has its risks, supporters of the nuclear industry point out that the environmental risk posed by coal-fired power stations is greater. There is the immediate risk posed by pollution through burning coal, and the long-term risk of global warming due to the emission of greenhouse gases.

In this section you will learn about the hazards of using radioactive materials.

▲ **Figure 4.19** The Fukushima Daiichi power plant in Japan.

TIP

In the early days of research into radioactivity, scientists did not understand the hazards of radiation. For example, Marie Curie (an early pioneer) died from an illness most likely caused by long term exposure to radiation. We now understand the impact of radiation because scientists have published their findings, which have been checked and validated by others. As a result we now have clear procedures to protect ourselves from the effect of radiation.

Radiation damage

Radiation entering our bodies causes damage in two ways.

- Direct damage is caused by a particle colliding with a cell in our body. An alpha particle, for example, behaves like a miniature bullet and it destroys body tissue.
- Indirect damage is caused by ionisation. Radiation produces ions which can make strong acids in our bodies. These acids can then destroy our cells or cause mutations in our genes.

Types of radiation

Inside the body the most damaging radiation is the alpha particle; this is because the alpha particle does not travel far through body tissues and it transfers its energy in a very small space. Therefore damage to tissue is localised.

Beta and gamma radiations spread their energy over greater distances in the body, so they are not as damaging as alpha radiation.

In nuclear reactors, neutrons are emitted. These particles are also dangerous to us, as they penetrate the body and collide with atoms, thus damaging cells.

Radiation dose

When we assess the dangers of radiation, we need to know how much we have been exposed to. We measure the radiation dose in **sieverts**, Sv. The greater the radiation dose, the more likely we are to suffer ill effects.

Doses are often quoted in thousandths of sieverts – millisieverts mSv.

TIP
The damage caused by an alpha particle inside our bodies is about 20 times the damage caused by beta or gamma radiation of the same energy.

How dangerous is radiation?

In the UK there is about a 40% chance that we will suffer from cancer at some stage of our life. It is thought possible that about 1% of cancers are caused by background radiation which sends thousands of particles through our bodies each second. The table below gives some examples of radiation doses and their risk to us.

Low doses

Low doses of radiation, below 10 mSv, are unlikely to cause us harm. However, it is possible that any exposure to radiation will increase our chances of cancer. We are exposed to background radiation all the time, in the air, from rocks and from the food we eat.

Moderate doses

Moderate doses of radiation below 1000 mSv (1 Sv) are unlikely to kill someone, but the person will be very unwell. Damage will be done to cells in the body, but not enough to be fatal. The body will be able to replace dead cells and the person is likely to recover completely. However, studies of the survivors from the Hiroshima and Nagasaki bombs, and of survivors from the Chernobyl reactor disaster of 1986, show that there is an increased chance of dying from cancer some years after the dose of radiation.

High doses

High doses of about 4000 mSv (4 Sv) are likely to be fatal, and a dose of 10 Sv will definitely be fatal. A high dose damages the gut and bone marrow so much that the body cannot work normally. About 30 people died of acute radiation syndrome in the Chernobyl disaster.

Radiation doses and their effects.

Dose/mSv	Source of dose	Effect of dose
0.0001	Airport security scan; Eating a banana	Low risk
0.005	Dental X-ray	Low risk
0.04	Transatlantic flight	Low risk
0.1	Chest X-ray	Low risk
0.2	Release limit from a nuclear plant per person per year	Low risk
0.4	Yearly dose from food	Low risk
2.4	Average from background radiation in UK	Low risk
3	Mammogram	Low risk
10	Average computer tomography (CT) scan	Low risk
50	Maximum yearly dose permitted for radiation workers	Medium risk
100	Lowest annual dose where increased risk of cancer is evident	Medium risk
1000	Highly targeted dose used in radiotherapy (single dose)	High risk, but balanced by a likely cure of cancer
5000	Extremely severe dose, received in a nuclear accident	Death probable within 6 weeks
10000	Maximum radiation dose per day found at the Fukushima plant in 2011	Fatal dose; death within 2 weeks
50000	10 minutes exposure to the Chernobyl reaction meltdown in 1986	Death within hours

Show you can...

Complete this task to show that you understand the hazards of radiation.

Explain how radiation can damage our bodies. Explain how measuring radiation doses can help reduce risk to people.

Test yourself

36 Name four nuclear radiations which are dangerous to us.

37 Explain why radiation workers are only allowed to receive a maximum annual dose of about 50 mSv.

38 A patient receives a dose of 10 mSv from a CT scan. Use the table on page 103 to work out how many years of background radiation is equivalent to the dose from a CT scan.

39 The table shows the results of research into the number of deaths caused by various types of radiation.

Source of radiation	Type of radiation	Number of people studied	Extra number of cancer deaths caused by radiation
Uranium miners	Alpha	3400	60
Radium luminisers	Alpha	800	50
Medical treatment	Alpha	4500	60
Hiroshima bomb	Gamma rays and neutrons	15000	100
Nagasaki bomb	Gamma rays	7000	20

a) Discuss whether the table supports the suggestion that alpha particles are more dangerous than gamma rays.

b) What conclusion can you draw about the relative dangers of neutrons?

c) A student comments that the table does not provide a fair test for comparison of the radiations. Comment on this.

Uses of radioactive materials

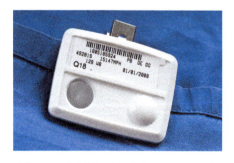

▲ **Figure 4.20** To check the amount of radiation that workers in a nuclear power station are exposed to, they wear special radiation-sensitive badges, like the one in this photograph. At the end of each month the sensitive film in the badges is developed and examined.

Radioactive materials have a great number of uses in medicine, industry and agriculture. People who work with radioactive materials must wear badges that record the amount of radiation to which they are exposed.

○ Radioactivity in medicine

Tracers

Some radioactive isotopes are used as tracers to help doctors examine the inside of our bodies. The isotope is attached to a biochemical agent which is then absorbed by an organ or area of the body to be examined. Technetium-99 is widely used as a tracer as it has several advantages.

● Technetium emits gamma rays which can be detected outside the body.

● The gamma rays are low energy so the damage to the body is minimised.

● The half-life of technetium-99 is 6 hours. This means doctors can detect high activity for scanning, but the radioactive material soon decays, making it safer for the patient (and people near to him or her).

Figure 4.21 shows an image of a baby's lungs which have absorbed a compound containing the tracer technetium-99. From this the doctor can see an abnormality in the left lung.

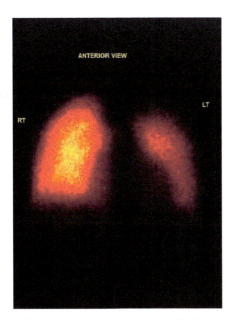

▲ **Figure 4.21** An image of a baby's lungs formed by detecting gamma rays emitted from the tracer technetium-99.

Tissue destruction

Certain cancers are treated by destroying them with high doses of radiation. This can be achieved, in some cases, by directing strong sources of gamma rays from outside the body. Other tumours can be destroyed by targeting then from inside the body with chemicals which have alpha or beta emitters inside them. Then short-range radiation supplies a high dose of radiation directly to the tumour.

⃝ Irradiation and contamination

When a teacher brings a radioactive source into a laboratory to demonstrate to her pupils, she irradiates the surroundings (with a very small safe dose) but does not cause any contamination. When the source is near a GM tube, the tube is irradiated. This means that radiation from the source is entering the tube. However, as soon as the source is removed and put away, the GM tube will not be radioactive, because there are no radioactive nuclei inside it to emit radiation.

Radioactive isotopes cause contamination when they get into places where we do not want them. For example, iodine-131 and caesium-137 were emitted in the Fukushima and Chernobyl disasters. Iodine-131 has a half-life of 8 days, which means that it is highly active for a short time. Iodine disperses in the atmosphere and gets into the food chain. This poses a very high risk, but for a short period. Caesium-137, however, has a half-life of 30 years. Consequently, the ground and water in the region close to the two nuclear reactors will be contaminated for many years to come.

Both irradiation and contamination are potentially hazardous to us. However, when a patient is exposed to a dose of radiation, for example, in hospital, the dose will be carefully calculated and controlled. The problem with contamination is that a person is exposed to radiation in a way which is unknown, and it is possible that a large and very dangerous dose of radiation is consumed by mistake.

Test yourself

40 Injecting a radioactive substance into a person to diagnose a medical problem always involves some risk to a person's health. Use an answer from the box to complete the sentence.

| greater than   the same as   less than |

A radioactive substance may be used to diagnose a medical problem if the potential benefit of the diagnosis is _____ than the risk to the person's health.

41 Explain what a medical tracer is.

42 State and explain two uses of radioactive isotopes in medicine.

43 a) Why is radioactive iodine very dangerous shortly after a nuclear accident?

b) Why does caesium-137 cause long-term problems of radioactive contamination? Why is contamination so dangerous to us?

44 Figure 4.22 shows a gamma ray unit in a hospital. Box A is a container of the radioactive material cobalt-60 ($^{60}_{27}$Co). This isotope has a half-life of 5 years.

a) Why is the container made mainly from lead?

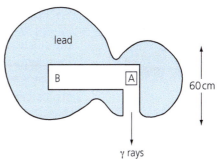

▲ **Figure 4.22**

b) The unit can be made 'safe' by moving the container to position B. Explain why.

c) How many protons, neutrons and electrons are there in one atom of cobalt-60?

d) The activity of the source was 2 million Bq in January 2016. What will it be in January 2026?

e) Suggest **three** safety precautions that someone should take when removing a container of cobalt-60 from the unit.

45 The table shows information about some radioisotopes. You are a medical physicist working in a hospital. Advise the doctors which isotopes are suitable for the following tasks. Explain your answers.

a) Checking a blockage in a patient's lung.

b) Targeting a tumour inside the body with alpha radiation therapy.

Isotope	Solid, liquid or gas at 20 °C	Type of radiation	Half-life
Hydrogen-3	Gas	Beta	12 years
Cobalt-60	Solid	Gamma	5 years
Strontium-90	Solid	Beta	28 years
Xenon-133	Gas	Gamma	5 days
Terbium-160	Solid	Beta	72 days
Actinium-227	Solid	Alpha	22 years
Bismuth-213	Solid	Alpha	45 minutes

Show you can...

Complete this task to show you understand how we make use of radioactive materials.

Explain two ways in which we use radioactive sources in hospitals.

Nuclear fission

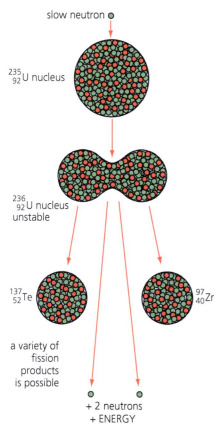

▲ **Figure 4.23** Nuclear fission.

The nuclei of some large atoms are unstable. To become more stable they lose an alpha or a beta particle. Some heavy nuclei, uranium-235 for example, may also increase their stability by the process of nuclear **fission**. Figure 4.23 shows how this works. Unlike alpha or beta decay, which happens at random, the fission of a nucleus is usually caused by a neutron hitting it. The $^{235}_{92}U$ nucleus absorbs this neutron and turns into a $^{236}_{92}U$ nucleus, which is so unstable that it splits into two smaller nuclei. The nuclei that are left are very rarely identical and two or three energetic neutrons are also emitted. The remaining nuclei are usually radioactive and will decay by the emission of beta particles to form more stable nuclei.

○ Nuclear energy

In Figure 4.23, just after fission has been completed, the tellurium-137 ($^{137}_{52}Te$) and the zirconium-97 nuclei ($^{97}_{40}Zr$) are pushed apart by the strong electrostatic repulsion of their nuclear charges. In this way the nuclear energy is transferred to the kinetic energy of the fission fragments. When these fast-moving fragments hit other atoms, this kinetic energy reduces and is transferred to the internal energy of the atoms causing a the temperature to rise.

The fission process releases a tremendous amount of energy. The fission of a nucleus provides about 40 times more energy than the release of an alpha particle from a nucleus. Fission is important because we can control the rate at which it happens, so that we can use the energy released to generate electricity in a nuclear power station.

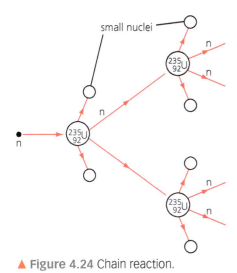

▲ **Figure 4.24** Chain reaction.

> **TIP** ✓
> You should be able to draw a chain reaction to answer exam questions.

⃝ Chain reaction

Once a nucleus has divided by fission, the neutrons that are emitted can strike other neighbouring nuclei and cause them to split as well. This **chain reaction** is shown in Figure 4.24. Depending on how we control this process, we have two completely different uses for it. In a controlled chain reaction, on average only one neutron from each fission will strike another nucleus and cause it to divide. This is what we want to happen in a power station.

In an uncontrolled chain reaction, most of the two or three neutrons from each fission strike other nuclei. This is how nuclear ('atomic') bombs are made. It is a frightening thought that a piece of pure uranium-235 the size of a tennis ball has enough stored energy to flatten a town.

⃝ Nuclear power stations

Figure 4.25 shows a gas-cooled **nuclear reactor**.

- The **fuel rods** are made of uranium-238, 'enriched' with about 3% uranium-235. Uranium-238 is the most common isotope of uranium, but it is only uranium-235 that will release energy by fission.
- The fuel rods are long and thin so that neutrons can escape. Neutrons leave one rod and cause another nucleus to split in a neighbouring rod.
- The rate of production of energy in the reactor is carefully regulated by the **boron control rods**. Boron absorbs neutrons very well, so by lowering the control rods into the reactor the reaction can be slowed down. In the event of an emergency, the rods are pushed right into the core of the reactor and the chain reaction stops completely.
- The neutrons emitted from the reactor are hazardous to humans. So a thick concrete shield is used to protect workers.

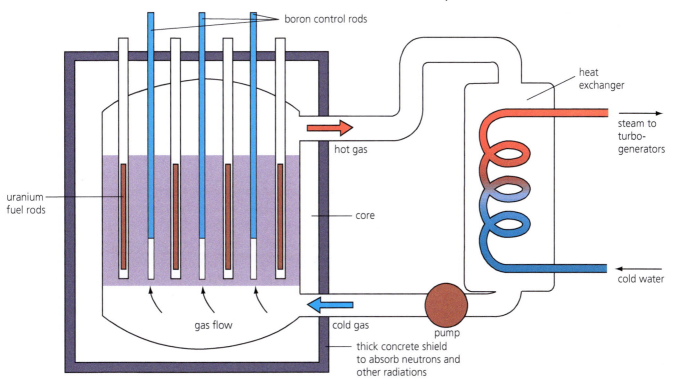

▲ **Figure 4.25** A nuclear reactor.

○ Nuclear equations

Figure 4.23 showed a possible fission of a uranium nucleus. This fission reaction can be described by an equation:

$$\,^{1}_{0}n + \,^{235}_{92}U \rightarrow \,^{137}_{52}Te + \,^{97}_{40}Zr + 2\,^{1}_{0}n$$

The mass and atomic numbers on each side balance. However, very accurate measurement shows that the mass on the left-hand side of the equation is slightly more than the mass on the right-hand side. This small reduction in mass is the source of the released nuclear energy.

○ Nuclear fusion

Two small nuclei may release energy when they join together to make a single larger nucleus. This is called **nuclear fusion**. This process is very difficult to achieve in controlled conditions, but occurs readily inside stars.

In a star two protons can fuse together to form a nucleus of deuterium $\,^{2}_{1}H$. This process releases energy as the mass of the deuterium nucleus is slightly less than the mass of two protons. This small amount of mass is converted to energy.

KEY TERM

Nuclear fusion is the name given to the process of joining small nuclei together to form larger ones.

Test yourself

46 What type of particle must the nucleus of a uranium-235 atom absorb before fission can happen?

electron   neutron   proton

47 What name is given to the process by which two small nuclei join together to form a single larger nucleus?

nuclear fission   nuclear fusion   nuclear reaction

48 Copy and complete the following sentences.
Both nuclear fusion and nuclear _____ reactions result in the release of _____. Nuclear fusion is the process by which energy is released in _____.

49 a) Explain what is meant by nuclear fission.
b) In what way is fission i) similar to, ii) different from radioactive decay?

50 What is a chain reaction? Explain how the chain reaction works in a nuclear bomb and in a nuclear power station.

51 Explain what is meant by the term nuclear fusion.

52 The following questions are about the nuclear reactor shown in Figure 4.25.
a) What is the purpose of the concrete shield surrounding the reactor?
b) Which isotope of uranium releases the energy in the fuel rods?
c) Will the fuel rods last forever? Explain your answer.
d) What would the operators do if the reactor core suddenly got too hot?

53 After absorbing a neutron, uranium-235 can split into the nuclei barium-141 ($\,^{141}_{56}Ba$) and krypton-92 ($\,^{92}_{36}Kr$). Write a balanced equation to show how many neutrons are emitted in this reaction.

Show you can...

Complete this task to show you understand the process of nuclear fission.

Explain the process of nuclear fission. Draw diagrams to illustrate a chain reaction.

Chapter review questions

1 Lithium atoms have three protons and four neutrons.

   a) What is the atomic number of lithium?

   b) What is the mass number of lithium?

   c) How many electrons are there in a lithium atom?

2 There are two stable isotopes of carbon: carbon-12 and carbon-13.

   a) Explain what is meant by the words

      i) stable

      ii) isotope.

   b) The atomic number of carbon is 6. How many protons and neutrons are there in each of the isotopes mentioned above?

3 How do the atoms of one particular element differ in atomic structure from the atoms of all other elements?

4 a) Write nuclear equations for the alpha decay of:

      i) $^{241}_{94}$Pu to uranium (U)

      ii) $^{229}_{90}$Th to radium (Ra)

      iii) $^{213}_{84}$Po to lead (Pb).

   b) Write nuclear equations for the beta decay of:

      i) $^{237}_{92}$U to neptunium (Np)

      ii) $^{59}_{26}$Fe to cobalt (Co)

      iii) $^{32}_{14}$Si to phosphorus (P).

5 The equations below are two examples of nuclear fission. Copy the equations and substitute numbers for the letters A to E.

   a) $^{1}_{0}n + ^{239}_{94}Pu \rightarrow ^{100}_{42}Mo + ^{134}_{A}Te + B^{1}_{0}n$

   b) $^{1}_{C}n + ^{235}_{92}U \rightarrow ^{140}_{54}Xe + ^{D}_{E}Sr + 2^{1}_{C}n$

6 Why did the work done by Geiger and Marsden convince scientists that the 'plum pudding' model of the atom needed to be replaced?

7 In the alpha scattering experiment about 1 in 10 000 alpha particles bounced back from the gold foil.

   Explain how you think the number of alpha particles bouncing back will change when:

   a) thicker gold foil is used

   b) aluminium foil of the same thickness is used.

8 Bismuth-213 emits alpha particles.

   a) What is an alpha particle?

   b) Explain why bismuth-213 would be highly dangerous if put inside the body.

9 Iodine-123 has a half-life of 13 hours. The isotope emits gamma rays. Explain why iodine-123 is used as a medical tracer.

10 Zak's teacher carried out an experiment to measure the half-life of protactinium-234. His results are shown in the table.

Time in seconds	Count rate in counts per second
0	66
40	44
80	30
120	20
160	13
200	9
240	6
280	4
320	4

a) In the experiment, what was:

  i) the independent variable

  ii) the dependent variable?

b) What instrument was used to measure the count rate?

c) What is the background count during the experiment?

d) Rewrite the table, subtracting the background count from each count rate value.

e) Plot a graph of the new count rate (vertically) against time (horizontally).

f) How long does it take for the count rate to fall from:

  i) 60 to 30

  ii) 40 to 20

  iii) 30 to 15?

g) Calculate an average value for the half-life of protactinium-234.

11 The table gives information about some radioactive sources that emit ionising radiation.

Source	Radiation emitted	Half-life
Bismuth-213	Alpha	45 minutes
Cobalt-60	Gamma	5 years
Uranium-233	Alpha	150000 years
Radon-226	Beta	6 minutes
Technetium-99	Gamma	6 hours

a) What is 'half-life'?

b) i) Explain what is meant by the term 'ionising radiation'.

  ii) Which of the radiations shown in the table is the most ionising?

c) Radiation has many uses in hospitals. Choose a source of radiation from the table for each of the following uses, explaining each of your choices:

  i) to sterilise plastic syringes sealed in a strong plastic bag

  ii) as a medical tracer that is injected into a body and then detected outside the body

  iii) to be used in a chemical to treat leukaemia inside the body.

Practice questions

1 The diagram represents an atom of beryllium-9.

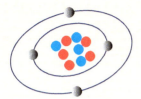

▲ Figure 4.26

a) i) Copy and complete the following table of information for an atom of beryllium-9. [3 marks]

Number of electrons	
Number of protons	
Number of neutrons	

ii) What is the atomic number of a beryllium-9 atom?

Choose the correct answer.

4 5 9 13 [2 marks]

Give the reason for your answer.

b) Beryllium-10 is a radioactive isotope of beryllium.

i) Choose the correct answer from the box to complete the sentence. [1 mark]

electron   neutron   proton

The nucleus of a beryllium-10 atom has one more _____ than the nucleus of a beryllium-9 atom.

ii) Beryllium-10 decays by emitting beta particles.

Which statement, A, B or C, describes a beta particle?

A the same as a helium nucleus

B an electromagnetic wave

C an electron from the nucleus [1 mark]

c) The graph shows how the count rate from a sample of beryllium-10 changes with time.

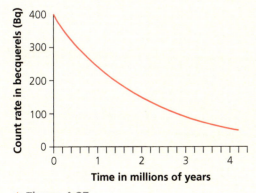

▲ Figure 4.27

i) How many millions of years does it take for the count rate to fall from 400 Bq to 40 Bq? [1 mark]

ii) What is the half-life of beryllium-10? [1 mark]

2 Figure 4.28 shows the average background radiation dose from natural sources that a person living in the UK receives in one year.

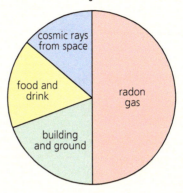

▲ Figure 4.28

a) i) The total radiation dose shown from all sources in Figure 4.28 is 2 millisievert (mSv).

What fraction of a sievert is a millisievert? [1 mark]

$\dfrac{1}{100}$ $\dfrac{1}{1000}$ $\dfrac{1}{1\,000\,000}$

ii) Work out the proportion of natural background radiation that comes from radon gas. [1 mark]

iii) Which one of the following, A, B or C is a correct conclusion from the data in Figure 4.28? [1 mark]

A Everyone receives the same dose from natural background radiation sources.

B Buildings and grounds produce more background radiation than cosmic rays.

C In the future the proportion of background radiation from cosmic rays will increase.

iv) Some background radiation is from man-made sources.

Name **one** source of background radiation that is man-made. [1 mark]

b) The table gives information about different radiation doses.

Radiation dose in mSv	
50000	Likely to cause death within hours
100	Lowest dose to cause an increased risk of cancer
50	Maximum dose allowed for radiation workers per year

A family goes on holiday for 2 weeks from the UK to Spain. The average background radiation dose in Spain is just over 5 mSv a year. This is more than double the average background radiation dose in the UK.

Explain why the family do not need to worry about the effect that the increased level of background radiation will have on their health. [3 marks]

3 a) The statements A, B and C give three properties of nuclear radiation.

A It will pass through cardboard but not thin aluminium.

B It is weakly ionising.

C It can travel only a few centimetres through the air.

i) Which **one** of the statements, A, B or C, gives a property of alpha radiation? [1 mark]

ii) Which **one** of the statements, A, B or C, gives a property of gamma radiation? [1 mark]

b) Fresh strawberries grown abroad are sometimes irradiated before being sent to the UK.

The irradiation process kills bacteria on the strawberries.

i) Which **one** of the statements, X, Y or Z, is correct? [1 mark]

X The irradiated strawberries do not become radioactive.

Y Particles containing radioactive atoms settle on the strawberries.

Z The strawberries cannot be eaten for a few days after irradiation.

ii) Suggest **one** reason why the farmers growing the strawberries want the strawberries to be irradiated. [1 mark]

4 a) Phosphorus is an element with an atomic number of 15. Its most common isotope is phosphorus-31, $^{31}_{15}P$. Another isotope, phosphorus-32, is radioactive.

i) State the number of protons, neutrons and electrons in phosphorus-31. [2 marks]

ii) Explain why phosphorus-31 has a different mass number from phosphorus-32. [1 mark]

iii) Atoms of phosphorus-32 change into atoms of sulfur by beta decay. Copy and complete the equation to show the atomic and mass numbers of this isotope of sulfur.

$$^{32}_{15}P \rightarrow ^{?}_{?}S + beta$$ [2 marks]

b) i) Name a suitable detector that could be used to show that phosphorus-32 gives out radiation. [1 mark]

ii) Name a disease that can be caused by too much exposure to a radioactive substance such as phosphorus-32. [1 mark]

5 A radioactive source emits alpha (α), beta (β) and gamma (γ) radiation.

a) Which **two** types of radiation will pass through a sheet of card? [1 mark]

b) Which type of radiation has the greatest range in air? [1 mark]

6 Several people who work in hospitals and in industry are exposed to different types of radiation. Three ways to check or reduce their exposure to radiation are:

A wear a badge containing photographic film

B wear a lead apron

C work behind a thick glass screen with remote-handling equipment.

Which method (A, B or C) should be used by the following people?

a) a hospital radiographer [1 mark]

b) a scientist experimenting with isotopes emitting gamma rays [1 mark]

c) a nuclear power station worker [1 mark]

7 a) Figure 4.29 shows what can happen when the nucleus of a uranium atom absorbs a neutron.

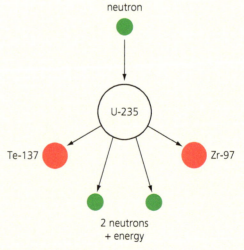

▲ Figure 4.29

i) What name is given to the process shown in Figure 4.29? [1 mark]

ii) Copy the diagram and add to it to show how this process could lead to a chain reaction. [3 marks]

iii) How does the mass number of an atom change when its nucleus absorbs a neutron?

| increases by 1   decreases by 1 |
| remains the same |

[1 mark]

8 The table gives information about some of the radioactive substances released into the air by the explosion at the Fukushima nuclear plant in 2011.

Radioactive substance	Half-life	Type of radiation emitted
Iodine-131	8 days	Beta and gamma
Caesium-134	2 years	Beta
Caesium-137	30 years	Beta

a) How is the structure of a caesium-134 atom different from the structure of a caesium-137 atom? [1 mark]

b) What are beta and gamma radiations? [2 marks]

c) A sample of soil is contaminated with some iodine-131. Its activity is 10 000 Bq. Calculate how long it will take for the activity to drop to 2500 Bq. [2 marks]

d) Which of the three isotopes will be the most dangerous 50 years after the accident? Explain your answer. [2 marks]

9 The table gives information about five radioactive isotopes.

Isotope	Radiation emitted	Half-life
A	Alpha	4 minutes
B	Gamma	5 years
C	Beta	12 years
D	Beta	28 years
E	Gamma	6 hours

a) What is a 'beta particle'? [1 mark]

b) What does the term 'half-life' mean? [1 mark]

c) Which of the isotopes could be used as a medical tracer? Explain the reason for your answer. [3 marks]

d) Radioactive waste needs to be stored. One suggestion is to seal it in steel drums and bury these in deep underground caverns. Suggest why people may be worried about having such a storage site close to where they live. [3 marks]

10 Some patients who suffer from cancer are given an injection of boron, which is absorbed by cancer cells. The cancerous tissue is then irradiated with neutrons. After this, the reaction that occurs is:

$$^{11}_{5}B \rightarrow ^{7}_{x}Li + ^{y}_{z}He$$

a) Copy the equation and write numbers in place of x, y and z. [3 marks]

b) Why do the lithium and helium nuclei repel each other? [1 mark]

c) Why can this treatment kill cancer cells? [1 mark]

d) Why is this process dangerous for healthy patients? [1 mark]

11 In the early twentieth century scientists thought that atoms were made up of electrons embedded inside a ball of positive charge. They called this the 'plum pudding' model of the atom.

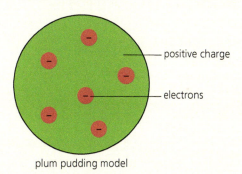

plum pudding model

▲ **Figure 4.30**

Geiger and Marsden fired a beam of alpha particles at a thin gold foil. Explain how the results of their experiment led to a new model of the atom. Illustrate your answer with suitable diagrams. [6 marks]

Working scientifically

Risk and perception of risk

Before starting an investigation you complete a risk assessment. You identify the hazards, the risks and the controls needed to reduce the identified risks. For example, if you use boiling water there is always a risk you may scald yourself.

1 Look at pages 96–98. What controls should be used to reduce the risk to the teacher and students of using a radioactive source?

Perception of risk

The perception of risk is often different from the actual risk. This happens for different reasons.

Voluntary versus imposed risk

When people are doing an activity by choice, they may perceive it to have a smaller risk than if they have no choice in doing something. For example, some people voluntarily throw themselves from high bridges and cranes, for example. They call it bungee jumping!

Familiar versus unfamiliar risk

People may be happy to use electrical appliances but worry about the risk of a nuclear reactor accident. In fact, many more people have been killed or injured by electrical accidents in the home than by accidents at nuclear reactors.

Visible versus invisible hazard

Every day we accept the risks involved in crossing the road. The hazards are real but our perception of the risk is likely to be reduced because generally we can see the hazards. Nuclear radiation, however, is invisible. We can't see it, we can't feel it, we can't smell it. The perceived risk is usually high.

2 Which of the following involves a voluntary risk?

   a) Flying to a holiday destination.

   b) Riding a bicycle to school.

   c) Exposure to background radiation.

   d) Learning to drive a car.

KEY TERM

Risk is the probability that something unpleasant will happen as the result of doing something.

▲ Figure 4.31 In a fall, a horse rider reduces the risk of serious injury by wearing a hard hat and body protector. Use the idea of momentum to explain how these help protect the rider from injury.

▲ Figure 4.32 For some people, having a suntan outweighs the risk of getting skin cancer from over exposure to ultraviolet radiation.

Electrical safety in the home

A recent survey found that many people perceive the risk of injury from using electrical appliances to be much lower than the accident figures suggest. On average, fires caused by the misuse of electrical appliances kill one person each week and seriously injure about 2500 people each year. In addition, thousands of people are injured each year (with about 30 deaths) by electric shocks. Perhaps it's because electrical appliances are so familiar to us that we simply forget or underestimate the risks.

Risk of using radioactivity in medicine

Most people would probably say that the risk to health posed by ionising radiation is high. Maybe this is because it is an invisible hazard with potential health risks that are difficult (if not impossible) for the individual to control.

On page 104 you will have read about the use of nuclear radiation for medical diagnosis and medical treatments.

Nuclear radiation causes ionisation. Inside our bodies, ionisation may damage cells. The cells may stop working properly and in some cases become a threat to the organism. Scientists know that a moderate to large dose of radiation increases the risk of developing a cancer. However, just as radiation can damage healthy cells, so it can be used to treat cancers.

The use of radiation in medical procedures is not without risk to the patient. In a group of patients all receiving a radiation dose of 0.1 Sv, some would experience health problems but for most they would not be long-lasting. However, why run the risk? Well it's not just a case of judging the risk in isolation; it must be considered in relation to the potential benefit.

▲ Figure 4.33 An overloaded socket increases the risk of a fire.

3 a) It is estimated that the risk of dying following a bone scan using technetium-99m is 1 in 5000. Why do patients take this risk?

   b) What distance do you think you would need to travel in a car to have the same risk of a serious accident?

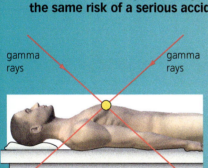

gamma rays   gamma rays

▲ Figure 4.34 The gamma rays from a cobalt-60 source are used to treat lung cancer. Each session, the gamma rays hit the cancer from a different direction so the cancer cells get a higher dose than the surrounding lung tissue.

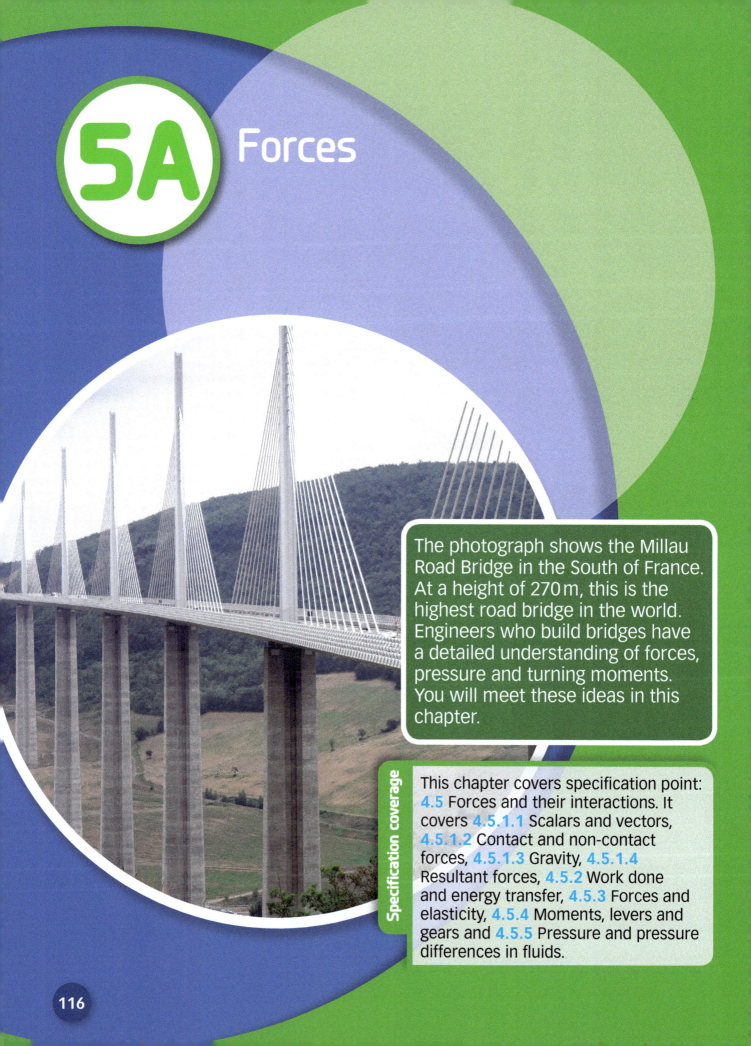

5A Forces

The photograph shows the Millau Road Bridge in the South of France. At a height of 270 m, this is the highest road bridge in the world. Engineers who build bridges have a detailed understanding of forces, pressure and turning moments. You will meet these ideas in this chapter.

Specification coverage

This chapter covers specification point: 4.5 Forces and their interactions. It covers 4.5.1.1 Scalars and vectors, 4.5.1.2 Contact and non-contact forces, 4.5.1.3 Gravity, 4.5.1.4 Resultant forces, 4.5.2 Work done and energy transfer, 4.5.3 Forces and elasticity, 4.5.4 Moments, levers and gears and 4.5.5 Pressure and pressure differences in fluids.

Previously you could have learned:

> A force is a push or a pull.
> A force can squash or stretch an object.
> A force can twist or turn an object.
> The resultant force acting on an object is the sum of the forces acting on the object.
> Turning moment = force × perpendicular distance from the pivot
> Pressure = $\dfrac{\text{force}}{\text{area}}$

Test yourself on prior knowledge

1 Name three common forces.
2 a) A force of 30 N is applied at right angles to a spanner of length 0.2 m to turn a nut. Calculate the turning moment exerted by the spanner on the nut.
   b) Why is a long spanner more useful than a short spanner when you need to tighten a nut?
3 A man with a weight of 900 N stands at rest on the ground. The area of his shoes is 0.06 m². Calculate the pressure his shoes exert on the ground.

Forces and their interactions

○ Scalars and vectors

A **scalar** quantity is one that only has a size or magnitude.

A **vector** quantity is one that has a size and a direction.

KEY TERMS

A **scalar** is a quantity with only magnitude (size) and no direction.

A **vector** is a quantity with both magnitude (size) and direction.

Some examples of scalar quantities are: mass (for example, 3 kg of apples); temperature (for example, 27 °C); energy (for example, 200 joules). These quantities do not have a direction, they only have a size.

Force is an example of a vector quantity. You can push with a force of 250 N to the right or with a force of 250 N to the left. Each force has the same magnitude, but the effect is in a different direction.

Velocity and speed

KEY TERM

Velocity is a speed in a defined direction.

Throughout this chapter, the words **velocity** and speed are both used to describe how fast something is moving. The difference between these two quantities is that speed is a scalar quantity and velocity is a vector quantity.

So when we say a car moves with a speed of 30 m/s we are just interested in how fast it is travelling. However, when we say that an aeroplane travels with a velocity of 180 m/s due east, we are saying two things: the direction the plane is travelling and how fast it is travelling.

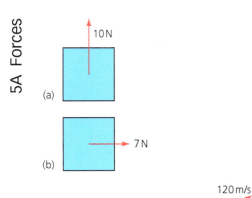

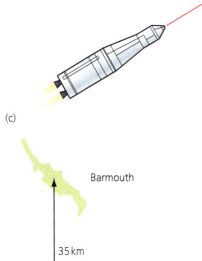

(c)

Barmouth

35 km

Abington

(d)

▲ **Figure 5.1** Examples of vectors.

Distance and displacement

You are used to describing a distance as a scalar quantity: for example, 'the ruler is 1 m long', or 'I ran 13 km this afternoon'. However, we are often interested in a direction too. For example, Bradford is 13 km due west of Leeds. When we add a direction to a distance, its correct name is a **displacement**.

Drawing vectors

You can represent a vector quantity by using an arrow. The length of the arrow represents the magnitude of the vector. The direction of the arrow shows the direction of the vector quantity. Some examples are given in Figure 5.1.

a) A force of 10 N acts upwards.

b) The box has a force of 7 N acting on it to the right. Note that the arrow here is shorter than the arrow in (a).

c) The rocket has a velocity of 120 m/s. The direction of the velocity tells us that the rocket is climbing as well as moving past a point on the ground.

d) When a car travels from Abington to Barmouth, its displacement is 35 km due north.

Test yourself

1 a) Which one of these is a scalar quantity?
   velocity   energy   displacement

  b) Which one of these is a vector quantity?
   mass   speed   force

2 What is wrong with this statement?
'The girl ran with a velocity of 8 m/s.'

3 a) A man walks 4 km due north. Make a drawing to show this displacement.

  b) The man in part (a) then walks 4 km due south. What is his displacement from his starting point? What distance did he travel?

  c) The next day the man in parts (a) and (b) walks 4 km east, then 3 km north.
   i) Draw a diagram to show his displacement from his starting point.
   ii) How far has the man walked?

  d) Explain the difference between distance and displacement.

Show you can...

Write a paragraph to explain the difference between a vector and a scalar quantity. Include some examples.

KEY TERMS

Displacement is a distance travelled in a defined direction.

A **force** is a push or a pull

⬭ **What is a force?**

A **force** is a push or a pull. Whenever you push or pull something you are exerting a force on it. The forces that you exert can cause three things:

- You can change the *shape* of an object. For example, you can stretch or squash a spring and you can bend or break a ruler.
- You can change the *speed* of an object. For example, you can increase the speed of a ball when you throw it and you decrease its speed when you catch it.
- A force can also change the *direction* in which something is travelling. For example, we use a steering wheel to turn a car.

KEY TERMS

A **contact force** can be exerted between two objects when they touch.

A **non-contact force** can sometimes be exerted between two objects that are physically separated.

▲ **Figure 5.2** The shot putter pushes the shot.

▲ **Figure 5.3** The archer pulls the string of the bow.

TIP

1 kN = 1 000 N (1 × 10^3 N)
1 MN = 1 000 000 N (1 × 10^6 N)

Contact forces

The forces described so far are called **contact forces**. Your hand touches something to exert a force.

Here are some further examples of contact forces.

- **Friction** is the contact force that opposes objects moving relative to each other. Friction acts between two sliding surfaces. Friction can also act to stop something beginning to move.
- **Air resistance**, or drag, is a force that acts on an object moving through the air. You can feel air resistance if you put your hand out of a car window when the car is moving. A boat experiences drag when it moves through water.
- **Tension** is the name we give to the force exerted through a rope when we pull something.
- **Normal contact force** is the force that supports an object that is resting on a surface such as a table or the floor.

Non-contact forces

There are also **non-contact forces**. These are forces that act between objects that are physically separated. Gravitational, electrostatic and magnetic forces are examples of non-contact forces.

Gravity

The Earth pulls you downwards whether you are in contact with it or not. The Sun's gravity acts over great distances to keep the Earth and other planets in orbit.

Electrostatic forces

These are forces that act between charged objects. By rubbing a balloon you can charge it and stick it to the wall.

Magnetic forces

A magnet can attract objects made from iron or steel towards it.

The size of forces

The unit we use to measure force is the **newton** (N). Large forces may be measured in kilonewtons (kN) or meganewtons (MN). A few examples of the size of various forces are given below.

The pull of gravity on a fly = 0.001 N

The pull of gravity on an apple = 1 N

The frictional force slowing a rolling football = 2 N

The force required to squash an egg = 50 N

The tension in a rope towing a car = 1000 N (1 kN)

The frictional force exerted by the brakes of a car = 5000 N (5 kN)

The push from the engines of a space rocket = 1 000 000 N (1 MN)

Example 1

Sakhib, who has a mass of 50 kilograms, weighs:

$$W = 50 \times 9.8 = 490 \text{ newtons}$$

Example 2

What is the weight of a 50 kg mass on the Moon's surface?

Answer

$$W = m \times g$$
$$= 50 \times 1.6$$
$$= 80 \text{ N}$$

▲ Figure 5.4 What is the weight of the apple?

KEY TERM

The **centre of mass** is the point through which the weight of an object can be taken to act.

KEY TERM

A number of forces acting on an object may be replaced by a single force that has the same effect as all the forces acting together. This single force is called the **resultant force**.

◯ Weight

Weight is the name that we give to the pull of gravity on an object. Large objects such as the Earth produce a gravitational field, which attracts masses. (Smaller objects exert gravitational forces on each other, but these are too small to notice.)

- Near the Earth's surface the gravitational field strength, g, is 9.8 N/kg.
- This means that each kilogram has a gravitational pull of 9.8 N.
- The weight of 1 kg on the Earth's surface is 9.8 N.

To calculate weight use this equation:

$$W = mg$$

weight = mass × gravitational field strength

Where weight is in newtons, N
mass is in kilograms, kg
gravitational field strength is in newtons per kilogram, N/kg.

Since the value of g is constant (9.8 N/kg near the Earth's surface) the weight of an object and the mass of an object are directly proportional.

weight ∝ mass

So if your mass in kilograms goes up or down, your weight in newtons will go up or down in the same proportion.

The mass of an object remains the same anywhere in the Universe. The mass of an object is determined by the amount of matter in it. However, a 50 kg mass would have a different weight on the Moon, where the gravitational field strength is 1.6 N/kg.

The weight of an object can be considered to act through a single point. This point is called the **centre of mass** of the object.

The weight of an object can be measured with a spring balance, which is calibrated in newtons.

◯ Resultant forces

A force is a vector, so we can represent the size and direction of a force with an arrow.

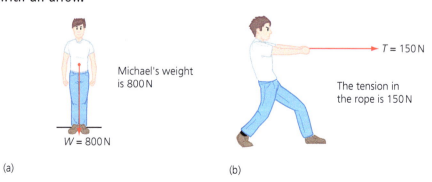

Michael's weight is 800 N

$W = 800$ N

(a)

$T = 150$ N

The tension in the rope is 150 N

(b)

▲ Figure 5.5 Two examples of forces acting on Michael: a) his weight (the pull of gravity on him) is 800 N. b) A rope with a tension of 150 N pulls him forwards.

TIP

The weight of an object depends on the gravitational field strength at the point where the object is.

When two forces act in the same direction, they add up to give a larger resultant force.

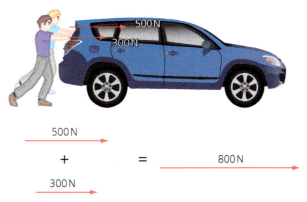

500 N

\+   =   800 N

300 N

▲ **Figure 5.6** Two people push the car, one with a force with of 300 N, and the other with a force of 500 N. The resultant force acting on the car is now 800 N to the right.

When two forces act in opposite directions, they produce a smaller resultant force. If the person pushing with 300 N in Figure 5.6 moves to the front of the car and exerts the same force, the resultant force will now be: 500 N – 300 N = 200 N to the right.

Resultant force and state of motion

Figure 5.7 shows two more examples of how forces add up along a line.

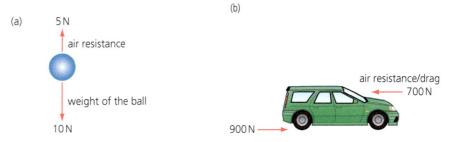

▲ **Figure 5.7** a) A ball is falling downward. The air resistance exerts an upwards force on the ball, and the weight of the ball exerts a downwards force on it. The resultant force of 5N speeds up the ball. b) A car is moving forwards and increases its speed. The push from the road is greater than the air resistance, so an unbalanced force of 200 N helps to speed up the car.

Test yourself

4 a) What is a contact force?
   b) Which one of the following is an example of a contact force?
       friction   gravity   magnetic
5 a) What is a non-contact force?
   b) Which one of the following is an example of a non-contact force?
       air resistance   electrostatic   tension
6 Explain how would you demonstrate that the magnetic force between two magnets is stronger than the gravitational pull on the magnets.
7 A girl has a mass of 57 kg. Calculate her weight on Earth.
8 Figure 5.8 shows some boxes. In each case, state what the resultant force is on the box. Give both the magnitude and direction of the force.

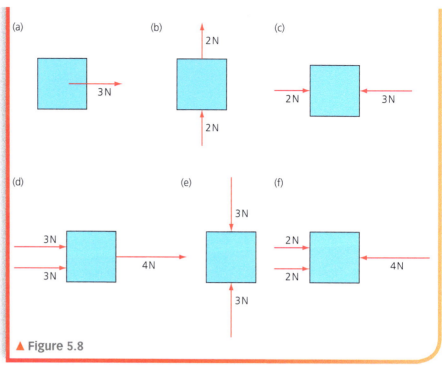

▲ Figure 5.8

Show you can...

Show you understand what a resultant force is by completing this task.

Draw vector diagrams to show how two forces, acting along the same line, can produce a) a larger resultant force, b) a smaller resultant force.

Free body diagrams

Free body diagrams are used to show the magnitude and direction of all of the forces that act on an object.

Figure 5.7a) is an example of a free body diagram. Two forces act on the ball: the air resistance exerts an upward force on the ball, and the pull of gravity (the ball's weight) exerts a downwards force on it. The sum of these two forces is 10 N – 5 N = 5 N downwards, so the ball continues to speed up.

Figure 5.9 shows a more complicated example of a free body diagram. Here, four forces are acting on Michael. Note that in drawing free body diagrams we treat the object as a point – so all the forces act through Michael's centre of gravity.

- In the vertical direction his weight, W, of 800 N is balanced by the normal contact force, R, from the floor of 800 N. So the resultant force upwards is zero.
- In the horizontal direction the pull from the rope, T, is 150 N and the frictional force, F, from the floor is 50 N. So the resultant horizontal force on Michael is 100 N to the right.

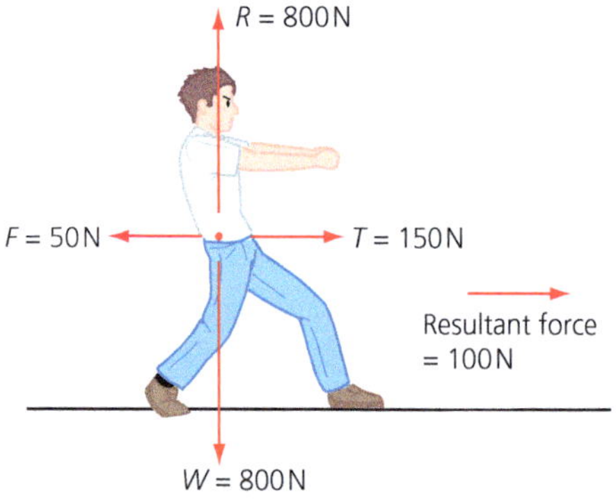

R = 800 N
F = 50 N   T = 150 N
Resultant force = 100 N
W = 800 N

▲ Figure 5.9 Forces acting on Michael.

Resolving forces

Figure 5.10a) shows a box which is being pulled along the floor, with a force of 10 N. However, the force is applied at an angle of 37° to the horizontal. How much of the force is being used to pull the box along the ground? The answer to the question is illustrated in Figure 5.10b).

Figure 5.10b) shows that the force can be resolved into two components, one vertical and one horizontal. The force is resolved in this way.

(a)

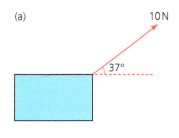

Draw the force to scale as shown by the red arrow in Figure 5.10a), at an angle of 37°.
Draw the horizontal component from the lower end of the red vector.
Draw the vertical component down from the higher end of the red vector to meet the horizontal component.

(b)

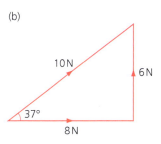

▲ Figure 5.10 a) A box is pulled along the floor at an angle. b) The pulling force on the box can be resolved into two components.

In this example the horizontal component of the force is 8 N and the vertical component is 6 N. The horizontal component of 8 N is used to overcome frictional forces exerted on the box by the floor, and the vertical component of the force acts to reduce the normal contact force between the box and the floor.

Test yourself

9 An astronaut has a mass of 120 kg in his spacesuit. Calculate his weight on Mars, where the gravitational field strength is 3.7 N/kg.

10 Draw a free body diagram for a tennis ball that is travelling mid-air in a horizontal direction. Include the drag acting on it.

11 Resolve each of the forces shown in Figure 5.11 into vertical and horizontal components.

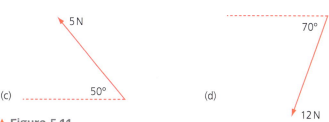

▲ Figure 5.11

TIP

Resolving a force.
A single force can be split into two components acting at right angles to each other. The two component forces together have the same effect as the single force.

▲ Figure 5.12

12 The box shown in Figure 5.10a) is stationary. It has a weight of 18 N.
a) State the frictional force on the box.
b) State the normal contact force acting on the box.
c) Draw a free body diagram to show the forces acting on the box.

13 Figure 5.12 shows a box on the ground; it is sliding to the left.
a) Draw a free body diagram to show the three forces acting on the box: friction, the box's weight and the normal contact force.
b) Use the free body diagram to explain why the box slows down.

Show you can...

Show that you understand what weight is by completing this task.

Explain why your weight would change if you moved to a different planet but your mass would remain the same.

TIP

Work = force applied × distance moved in direction of force

▲ **Figure 5.13** This man does work when he lifts the wheelbarrow, and when he pushes it forwards.

TIP

Large amounts of work are measured in kJ (1000 J or 10^3 J) and MJ (1 000 000 J or 10^6 J).

○ Work done and energy transfer

A job of work

Tony works in a supermarket. His job is to fill up shelves when they are empty. When Tony lifts up tins to put them on the shelves, he is doing some work. The amount of work Tony does depends on how far he lifts the tins and how heavy they are.

The amount of work done can be calculated using the equation:

$$W = F\,s$$

work done = force × distance moved in the direction of the force

Where work done is in joules, J
force is in newtons, N
distance is in metres, m

Work is measured in **joules** (J). One joule of work is done when a force of 1 newton moves something through a distance of 1 metre, in the direction of the applied force.

$1\,J = 1\,N \times 1\,m$

so, 1 joule = 1 newton-metre

> **Example**
>
> How much work does Tony do when he lifts a tin weighing 20 N through a height of 0.5 m?
>
> **Answer**
>
> $W = F \times s$
> $= 20 \times 0.5$
> $= 10\,J$
>
> Tony would do the same amount of work if he lifted a tin weighing 10 N through 1 m.

You will notice that the unit of work, J, is the same as that used for energy. This is because when work is done, energy is transferred.

- In the case of the tins, the work done by Tony is transferred to the tins as gravitational potential energy.
- If you do work dragging a box along the ground, you are working against frictional forces. This causes the temperature of the box and the ground to rise.
- If you push start a car, the work you do is transferred to the kinetic energy of the car.

Does a force always do work?

A force does not always do work. In Figure 5.14, Martin is helping Salim and Teresa to give the car a push start. Teresa and Salim are pushing from behind and Martin is pushing from the side. Teresa and Salim are doing some work because they are pushing in the right direction to get the car moving. Martin is doing nothing useful to get the car moving. Martin does no work because he is pushing at right angles to the direction of movement.

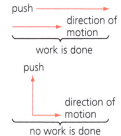

▶ **Figure 5.14** Teresa and Salim are doing work but Martin is not.

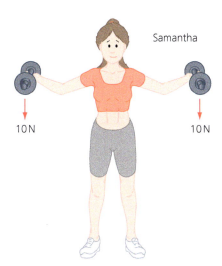

▲ **Figure 5.15** Samantha is not doing any work because the weights are not moving.

In Figure 5.15, Samantha is doing some weight training. She is holding two weights but she is not lifting them. She becomes tired because her muscles transfer energy, but she is not doing any work because the weights are not moving. To do work you have to move something, for instance lifting a suitcase or pushing a car along the road. When a car is idling at traffic lights, energy is transferred, but no work is done against resistive forces. The car engine only does work when the car is moving.

Test yourself

14 In which of the following cases is work being done? Explain your answers.
   a) A magnetic force holds a 'fridge magnet' on a steel door.
   b) You pedal a bicycle along a road.
   c) A crane is used to lift a large bag of sand.
   d) You hold a bag of shopping without moving it.
15 Calculate the work done in each case.
   a) You lift a 20 N weight through a height of 2.5 m.
   b) You drag a box 8 m along a floor using a force of 75 N, parallel to the ground.
16 Copy and complete the table.

Task of work	Force applied in newtons	Distance moved in metres	Work done in joules
Opening a door	9 N	0.8 m	
Pulling a wheeled suitcase		125 m	2500 J
Lifting a box of shopping	150 N		225 J
Pushing a toy	5 N	7 cm	
Driving a car along the road	750 N	20 km	

17 Joel is on the Moon in his spacesuit. His mass, including his spacesuit, is 150 kg. The gravitational field strength on the Moon is 1.6 N/kg.
   a) Calculate Joel's weight.
   b) Joel climbs an 8 m ladder into his spacecraft. How much work has he done?

Show you can...

Show you understand what work is by completing this task.

Write down an equation which links work, force and distance. Explain why when a force is applied to an object, work is not always done.

▲ **Figure 5.16** An unbalanced force will change the speed of a car.

◯ Forces and elasticity

If one force only is applied to an object, for example a car then it will change speed or direction. If we want to change the shape of an object, we have to apply more than one force to it.

Figure 5.17 shows some examples of how balanced forces can change the shape of some objects. Because the forces balance, the objects remain stationary.

- Two balanced forces can stretch a spring.
- Two balanced forces can compress a beam.
- Three balanced forces cause a beam to bend.

(a)

(b)

(c)

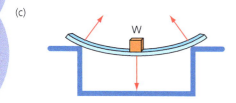

▲ **Figure 5.17** Balanced forces can change the shape of some objects.

Sometimes when an object has been stretched, it returns to its original length after the forces are removed. If this happens, the object experiences **elastic deformation**.

Sometimes an object that has been stretched does not return to its original length when the forces are removed. If the object remains permanently stretched, the object experiences **inelastic deformation**.

Elastic and inelastic deformations can be shown easily by stretching a spring in the laboratory. When small forces are applied and then removed, the spring returns to its original length and shape. When large forces are applied and then removed, the spring does not return to its original length.

You can also explore elastic and inelastic behaviour with an empty drinks can. When you squeeze the can gently, it springs back to its original shape when you remove your fingers. However, by applying larger forces you can change the can's shape permanently.

Stretching a spring

For a spring that is elastically deformed, the force exerted on a spring and the **extension** of the spring are linked by the equation:

$$F = k \times e$$

force = spring constant × extension

Where: force is in newtons, N
spring constant is in newtons per metre, N/m
extension is in metres, m

The spring constant is a measure of how stiff a spring is. If k is large, the spring is stiff and difficult to stretch. When a spring has a spring constant of 180 N/m, this means that a force of 180 N must be applied to stretch the spring 1 m. The equation, $F = k \times e$, can also be applied to the compression of a stiff spring by two forces. In this case, e is the distance by which the spring has been compressed (squashed).

The work done in stretching or compressing a spring (up to the limit of proportionality see p. 128) can be calculated using the equation:

$$\text{elastic potential energy} = \frac{1}{2} \times \text{spring constant} \times (\text{extension})^2$$

$$E_e = \frac{1}{2} k e^2$$

More details and an example are on pages 5 and 6.

KEY TERMS

Elastic deformation occurs when an object returns to its original length after it has been stretched.

Inelastic deformation occurs when an object does not return to its original length after it has been stretched.

Extension is the difference between the stretched and unstretched lengths of a spring.

Example

George uses a spring to weigh a fish he has just caught. The spring stretches 8 cm. George knows that the spring constant is 300 N/m. Calculate the weight of the fish.

Answer

$$F = k \times e$$
$$= 300 \times 0.08$$
$$= 24\,\text{N}$$

Remember: you must convert the 8 cm to 0.08 m, because the spring constant is measured in N/m.

Investigating the relationship between force and the extension of a spring

How much a spring stretches depends on the force applied to the spring. A 100 g mass hung from the end of a spring exerts a force of 1 newton on the spring.

You can investigate the extension of a spring using the apparatus shown in Figure 5.18. During the investigation you should wear a pair of safety glasses.

Method

1. Set up the retort stand, metre ruler and steel spring as shown in Figure 5.18. Make sure that the metre ruler is vertical.
2. Clamp the retort stand to the bench.
3. Measure the position of the bottom of the spring on the metre ruler, l_1.
4. Draw a suitable table to record the force acting on the spring, the metre ruler readings and the calculated values for the extension of the spring.
5. The first result to record is when no mass is attached to the spring, the force applied to the spring is then 0 and the extension of the spring is 0.
6. Hang a 100 g slotted mass hanger from the bottom of the spring. The force exerted on the spring by the mass hanger is 1 N.
7. Measure the new position of the bottom of the spring, l_2.
8. Calculate the extension of the spring: $e = l_2 - l_1$
9. Add a 100 g mass to the hanger. The total mass is now 200 g so the force exerted on the spring is 2 N.
10. Measure the new position of the bottom of the spring, l_3, and calculate the total extension ($l_3 - l_1$).
11. Add a 3rd, 4th, 5th and 6th 100 g mass to the hanger. Each time you add a mass, measure the new position of the bottom of the spring and calculate the total extension.

Analysing the results

1. Plot a graph of the extension of the spring against the force applied to stretch it.
2. Draw a straight line of best fit through the plotted points.

   Your graph should show the same pattern as the graph drawn in Figure 5.19.
3. What can you conclude from this investigation?

Taking it further

Using the same method, you could find out whether the extension of a rubber band follows the same pattern as a steel spring.

Questions

1. What precautions should be taken when carrying out this investigation to reduce the risk of injury?
2. Was the method given for this investigation a **valid** method? Give a reason for your answer.
3. A second set of results could have been taken without having to start again and repeat the investigation. Suggest how.
4. Suggest one way that the accuracy of the measurements taken during the investigation could have been improved.

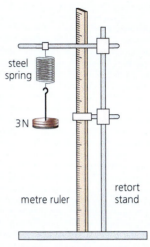

▲ **Figure 5.18** Apparatus for the investigation.

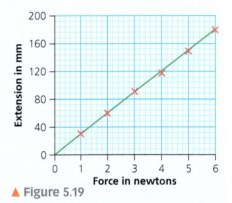

▲ **Figure 5.19**

KEY TERM

The method used in an investigation is **valid** if it produces data that answers the question being asked.

KEY TERM

A spring is permanently deformed when it is extended beyond the **limit of proportionality**.

Limit of proportionality

The graph in Figure 5.20 shows what happens if you put enough weights on the spring to cause inelastic deformation. There comes a point when the spring is permanently deformed, and it has gone beyond the **limit of proportionality**. Now when the weights are removed, the spring does not return to its original length.

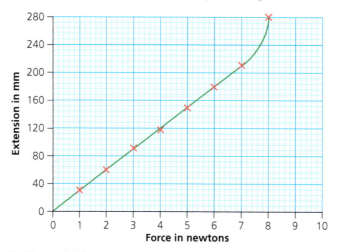

▶ Figure 5.20

The spring constant for the spring can be calculated from the linear section of the graph by dividing the force by the extension:

$$k = \frac{F}{e}$$

Elastic potential energy

When a spring is deformed elastically, energy is stored in the spring as **elastic potential energy.** The further the spring is stretched, the greater the amount of elastic potential energy stored in the spring.

However, if an object is deformed inelastically, there is no elastic potential energy stored. For example, when you crush a drinks can, work has been done moving atoms past each other to make a new shape. The work done causes a small temperature rise in the can.

Test yourself

18 Each of the objects in Figure 5.21 has had its shape changed. Which of the objects has stored elastic potential energy? Give a reason for each of your answers.

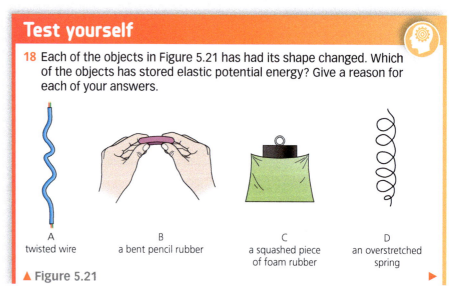

| A | B | C | D |
| twisted wire | a bent pencil rubber | a squashed piece of foam rubber | an overstretched spring |

▲ Figure 5.21

19 Explain what is meant by elastic deformation.
20 a) An object with an unknown weight is attached to the spring used in Figure 5.18. It causes the spring to extend by 136 mm. Use the graph in Figure 5.19 to calculate the weight of the object.
   b) Use the graph in Figure 5.19 to calculate the spring constant for the spring.
21 A force of 600 N causes the suspension spring of a car (Figure 5.22) to compress by 3 cm. Calculate the spring constant of the spring.

▲ Figure 5.22

22 Describe an experiment which you would use to determine the spring constant of a spring.

Show you can...

Show you understand what is meant by inelastic deformation by completing this task.

Design an experiment to show that a sheet of aluminium can be stretched both elastically and inelastically. Be careful to use exact language to describe your plan.

▲ Figure 5.23 A long spanner is used to undo the nuts on the wheel of a car.

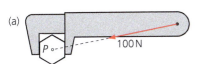

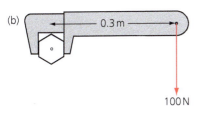

▲ Figure 5.24 Why perpendicular distance is important.

Moments, levers and gears

If you have ever tried changing the wheel of a car, you will know that you are not strong enough to undo the nuts with your fingers. You need a spanner to get a large **turning effect**. You need a long spanner and a large force.

The turning effect of a force about a pivot is called the **moment** of the force. The moment can be increased by applying a large force that is a long distance from the pivot.

The size of the moment is defined by the equation:

$$M = F \times d$$

moment of a force = force × distance

Where moment of a force is measured in newton-metres, Nm
force is in newtons, N
distance, d, is the perpendicular distance from the pivot to the line of action of the force, in metres, m.

Perpendicular means 'at right angles'. Figure 5.24 shows you why this distance is important. In Figure 5.24a) you get no turning effect at all as the force acts straight through the pivot, P. In Figure 5.24b) you can calculate the turning moment.

turning moment = 100 N × 0.3 m
             = 30 Nm

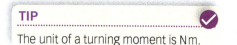
Lifting loads

The turning effect of a force needs to be considered when a heavy load is lifted with a mobile crane (Figure 5.25). If the turning effect of the load is too large then the crane will tip over. So, inside his cab, the crane operator has a table to tell him the greatest load that the crane can lift for a particular working radius.

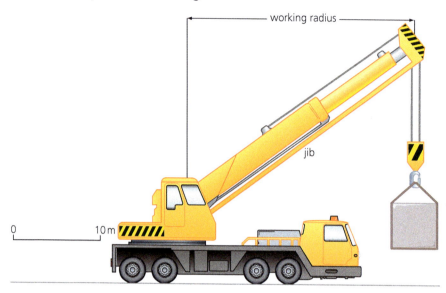

▲ **Figure 5.25** A mobile crane.

The table below shows you how this works. For example, the crane can lift a load of 600 kN safely with a working radius of 16 m. If the crane is working at a radius of 32 m, it can only lift 300 KN. You get the same turning effect by doubling the working radius and lifting half the load.

A load table for a crane operator

Working radius in m	Maximum safe load in kN	Load × radius/kN × m
12	800	9600
16	600	9600
20	480	9600
24	400	9600
28	340	9520
32	300	9600
36	265	9540

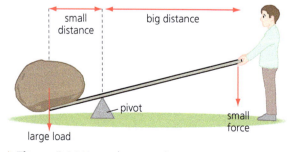

▲ **Figure 5.26** How a lever works.

Levers

We can use the idea of turning moments to make life easier. Figure 5.26 shows how a lever works. The principle is that the turning moment applied by the man on the right-hand side of the pivot is transmitted to the rock on the left-hand side of the pivot. So the small force exerted by the man with the long lever applies a large force to the rock that is a small distance from the pivot.

Example

In Figure 5.26 the man applies a downwards force of 150 N at a big distance of 2.0 m from the pivot. A rock is placed a small distance of 0.4 m from the pivot.

How much force does the lever exert on the rock?

Answer

turning moment exerted by the man = turning moment exerted on the rock

So $150 \times 2 = F \times 0.4$

$F = 750\,N$

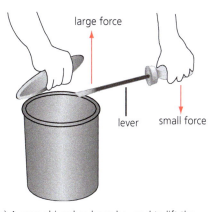

a) A screwdriver has been be used to lift the lid off a tin of paint.

b) A screwdriver is used to insert a screw – a wide handle allows us a good grip and leverage.

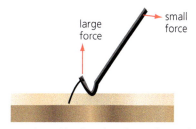

c) A crowbar with a long handle can be used to pull out old nails.

▲ **Figure 5.27** Further examples of levers.

Balancing

In Figure 5.28 you can see Jaipal and Mandy sitting on a balanced seesaw. It is balanced because the turning moment produced by Jaipal's weight exactly balances the turning moment produced by Mandy's weight in the opposite direction.

Jaipal's anti-clockwise turning moment $= 450 \times 1$

$= 450\,Nm$

Mandy's clockwise turning moment $= 300 \times 1.5$

$= 450\,Nm$

The turning moments balance. The seesaw is in equilibrium (balanced).

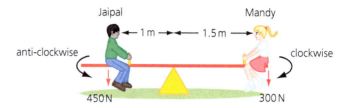

▲ Figure 5.28 A balanced seesaw.

KEY TERM

The **principle of moments** states that when a system is balanced the sum of the anti-clockwise turning moments equals the sum of the clockwise turning moments.

In equilibrium the sum of the anti-clockwise moments = the sum of the clockwise moments. This is called the **principle of moments**.

Example

Mandy gets off the seesaw, and Claire gets on and sits in the same place. Claire weighs more than Mandy. Which way should Jaipal move to balance the seesaw?

Answer

Claire's weight produces a larger turning moment than Mandy, so Jaipal needs to move further away from the pivot, so his weight produces a larger turning moment.

Moments in action

In Figure 5.29, a small crane is being used to load a boat. A counterbalance weight on the left-hand side is used to stabilise the crane and stop it from toppling over.

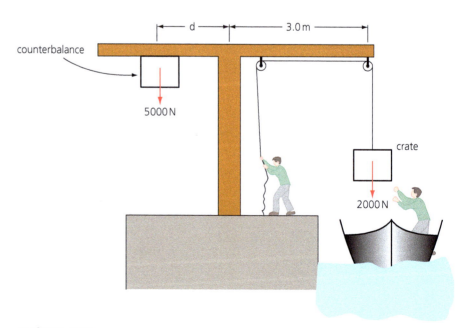

▲ Figure 5.29

Example

Where should the counterbalance weight be placed when the crew use the crane to lift a 2000 N crate into the boat?

Answer

The anticlockwise moment   The clockwise moment
from the counter balance = from the crate

$$5000 \times d = 2000 \times 3.0$$
$$d = \frac{6000}{5000}$$
$$d = 1.2\,m$$

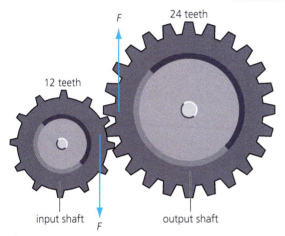

▲ **Figure 5.30** Where gear teeth touch, equal and opposite forces act. The output gear wheel has twice the radius of the input wheel.

Gears

Many machines make use of **gears**. Gears can be used to increase or decrease the rotational effects of a force. Figure 5.30 shows how a machine can produce a large turning effect.

The turning effect from the input shaft $= F \times r$
The turning effect from the output shaft $= F \times 2r$

The teeth of the input and output wheels exert the same force, F, on each other. However, because the output wheel has twice the radius of the input wheel, the turning effect from the output shaft is twice as big as the turning effect applied by the input shaft.

Test yourself

23 Each drawing in Figure 5.31 shows a force that has a turning effect. Calculate the moment of each force about the pivot, P.

a)

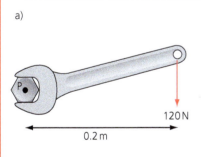

120 N

0.2 m

b)

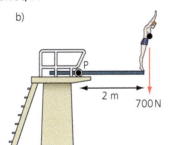

2 m

700 N

c)

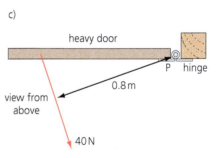

heavy door

0.8 m

P   hinge

view from above

40 N

▲ **Figure 5.31**

24 Explain the following.
   a) A mechanic chooses a spanner with a long handle to undo a tight nut.
   b) The handle of a door is placed a long way from the hinge.
25 a) Why do cranes have counterbalances?
   b) In Figure 5.29, the crew want to load a heavier crate into the boat. Which way should they move the counterbalance to balance the crane? Explain your answer by referring to turning moments.

c) When the load in Figure 5.29 is 2500 N, the crew move the counterbalance to a distance of 1.5 m.
   i)   Calculate the turning moment of the load.
   ii) Calculate the turning moment of the counterbalance.
   iii) Explain why the crane balances.

26 A student finds that the minimum force required to lift up the lid of a computer is 6.4 N (Figure 5.32).
   a) Calculate the moment of the force that opens the computer.
   b) Explain why the minimum force required to close the computer lid is likely to be less than 6.4 N (Figure 5.32).

27 Look at the crane in Figure 5.25.
   a) What is meant by the term 'working radius'?
   b) Use the scale to calculate the working radius of the crane in this position.
   c) Use the data in the table on page 130 to calculate the greatest load that the crane can lift safely in this position.

28 The diagram shows a pair of cutters.

6.4 N
0.25 m
P

▲ Figure 5.32

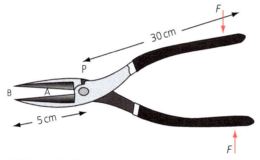

F
30 cm
P
B A
5 cm
F

▲ Figure 5.33

a) Explain why the cutters have long handles.
b) You want to cut some thick wire. Use your knowledge of moments to explain whether it is better to put the wire at point A or point B.
c) A wire that needs a force of 210 N to cut it is placed at B, 5 cm from the pivot. Calculate the size of the force you need to apply at the ends of the handles.

29 In Figure 5.34 the teeth of gear wheels exert a force of 250 N on each other. The radius of the smaller wheel is 10 cm, and the larger wheel has a radius of 20 cm.
   a) Calculate the turning moment produced by each wheel.
   b) Explain how gears can be used in machines.
   c) Which gear wheel is likely to wear out first?

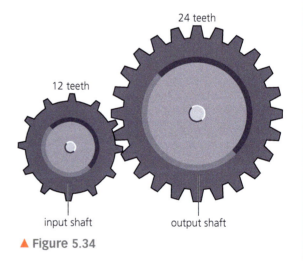

24 teeth
12 teeth
input shaft
output shaft

▲ Figure 5.34

Show you can...

Show that you have understood the principle of moments by completing this task.

Draw a horizontal see-saw with two children sitting on it. Explain what condition must be met for the see-saw to balance. How does this work with three or more children on the see-saw?

Pressure and pressure differences in fluids

KEY TERM

A **fluid** is a liquid or a gas. A fluid flows and can change shape to fill any container.

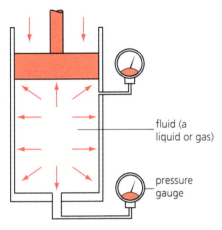

▲ **Figure 5.35** The pressure in a fluid acts equally in all directions.

○ Pressure in a fluid

When a ruler is used to push a lot of marbles lying on a table, they do not all move along the direction of the push. Some of the marbles give others a sideways push. The marbles are behaving like a **fluid**. The name fluid is applied to something that flows. Fluids are liquids or gases. The marbles are modelling the behaviour of the particles in the fluid.

In Figure 5.35, you can see a cylinder of fluid that has been squashed by pushing a piston down. The pressure increases everywhere in the fluid, not just next to the piston. The fluid is made up of lots of tiny particles that act rather like the marbles to transmit the pressure to all points.

The pressure exerted by the fluid acts equally in all directions, and acts at right angles to the surface of the container walls. We calculate the pressure exerted by a force using the equation:

$$P = \frac{F}{A}$$

$$\text{pressure} = \frac{\text{force normal to a surface}}{\text{area of that surface}}$$

where pressure is in pascals, Pa

force is in newtons, N

area is in metres squared, m^2.

In this equation the word 'normal' means at right angles to the surface.

Hydraulic machines

We often use liquids to transmit pressures. Liquids can change shape but they hardly change their volume when compressed. Figure 5.36 shows how a hydraulic jack works.

A force of 50 N presses down on the surface above A. The extra pressure that this force produces in the liquid is:

$$P = \frac{F}{A}$$
$$= \frac{50}{10} = 5 \, \text{N/cm}^2$$

The same pressure is passed through the liquid to B. So the upwards force that the surface above B can provide is:

$$F = P \times A$$
$$= 5 \times 100 = 500 \, \text{N}$$

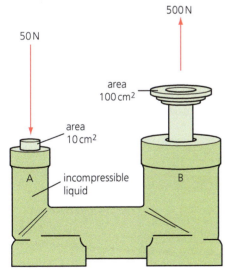

▲ **Figure 5.36** How a hydraulic jack works.

With the hydraulic jack you can lift a load of 500 N by applying a force of only 50 N. Figure 5.37 shows another use of hydraulics.

▶ **Figure 5.37** Cars use a hydraulic braking system. The foot exerts a small force on the brake pedal. The pressure that this force creates is transmitted by the brake fluid to the brake pads. The brake pads have a large area and exert a large force on the wheel disc. The same pressure can be transmitted to all four wheels.

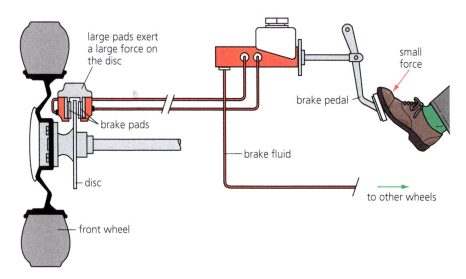

Increase of pressure with depth

The scuba diver in Figure 5.38 is enjoying warm tropical waters, where she can see brightly coloured fish in coral reefs. The water looks clear and safe, but one danger is the pressure of the water itself.

The pressure that acts on a diver depends on the weight of water above her. As she goes deeper, the height and weight of the water above her increase. This increases the pressure on the diver.

For every 10 m below the surface of the sea, there is an increase of about 1 atmosphere pressure. This diver is experiencing a pressure of 3 atmospheres at a depth of 20 m.

The diver feels a bigger pressure under 10 m of sea water than under 10 m of fresh water. This is because sea water has a higher density than fresh water. So a column of sea water would have a greater mass and therefore greater weight than an identical column of fresh water.

▲ **Figure 5.38** This diver experiences a pressure of 3 atmospheres at a depth of 20 m.

The pressure at the bottom of a column of liquid depends on three things:

- height of the column, h
- density of the liquid, ρ
- gravitational field strength, g (g = 9.8 N/kg).

The pressure due to a column of liquid can be calculated using the equation:

$$P = h\,\rho\,g$$

pressure = height of the column × density of the liquid
× gravitational field strength

where pressure is in pascals, Pa

height is in metres, m

density is in kilograms per metre cubed, kg/m³

gravitational field strength is in newtons per kilogram, N/kg.

Example

What pressure does a diver experience at a depth of 16m in a diving tank? The density of water is 1000 kg/m³ and atmospheric pressure is 100 000 Pa.

Answer

The extra pressure due to the water is:

$$P = h \rho g$$
$$= 16 \times 1000 \times 9.8$$
$$= 156\,800\,\text{Pa}$$

So the total pressure is 100 000 Pa + 156 800 Pa = 256 800 Pa.

Floating

Figure 5.39 shows a metal cylinder that is submerged in water. The arrows represent the size of the pressure acting around the cylinder. Because the pressure exerted by the water increases with depth, the pressure on the bottom of the cylinder is larger than the pressure acting on the top of the cylinder. This difference in pressure creates a resultant force upwards.

The resultant force upwards exerted by a fluid on an object is called the **upthrust**.

The size of the upthrust is equal to the weight of the fluid displaced by the object.

An object floats if the upthrust from a fluid is equal to the object's own weight. If the weight of the object is greater than the upthrust, the object sinks.

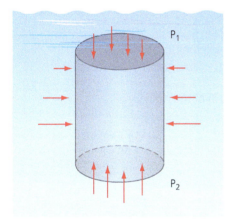

▲ **Figure 5.39** There is a resultant force upwards because there is a difference in pressure between the top and bottom of the cylinder.

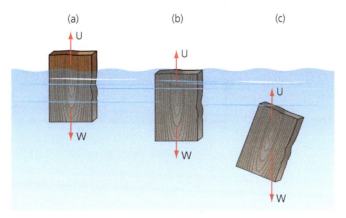

▲ **Figure 5.40** Three blocks of wood in water. They have different densities.

- Block (a) has a density lower than that of water. It displaces a weight of water equal to its own weight. So it floats.
- Block (b) has a density equal to that of water. It displaces a weight of water equal to its own weight, when it is fully submerged.
- Block (c) has a density greater than that of water. Now the weight is greater than the upthrust so it sinks.

Test yourself

30 Explain why the pressure under the surface of the sea increases with depth.

31 Sometimes, after a road accident, the Fire and Rescue Service uses inflated air bags to lift a vehicle, to free passengers who have become trapped. Explain how such bags can lift a large load easily.

32 a) Calculate the pressure acting on a diver at depths of i) 5 m, ii) 28 m. The density of the water is 1000 kg/m³. You will need to include the pressure of the atmosphere at the surface of the sea, which is 100 000 Pa.

b) Calculate the force exerted by the pressure on a diver's mask at a depth of 30 m. The mask has an area of 0.02 m².

c) Explain why the diver's mask is in no danger of breaking under this pressure.

33 A long cylinder hangs from a spring balance. As the cylinder is lowered into some water, measurements are taken from the spring balance for difference values of d (see Figure 5.41 and the table).

a) Plot a graph of the balance reading (y-axis) against the distance d (x-axis).

b) Use your graph to work out the length of the cylinder.

c) What is the weight of the cylinder?

d) When the cylinder is fully submerged, what is the upthrust on it?

e) Draw a free body diagram to show the forces acting on the cylinder when it is fully submerged.

d in cm	Balance in N
0	51
10	46
20	41
30	36
40	32
50	32
60	32

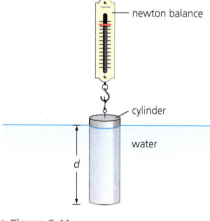

▲ Figure 5.41

Show you can...

Show you understand why some objects float by completing this task.

Explain why a fluid causes an upthrust on a submerged or partially submerged object. What determines whether an object floats?

Atmospheric pressure

One of the most effective ways of moving pieces of glass on a building site is to pick them up using large suction pads. The idea is simple: you remove the air from inside the pad and atmospheric pressure pushes the pad firmly in place against the glass.

▲ **Figure 5.42** Large suction pads can be used to move large pieces of glass.

Changing atmospheric pressure

Atmospheric pressure is caused by billions of air molecules colliding with a surface each second. At room temperature, air molecules are travelling with speeds of about 500 m/s. All these fast collisions add up to produce a large force on an object.

The pressure exerted by a gas depends on the number of molecules present in a volume of that gas. Near the Earth's surface, the pull of gravity compresses the atmosphere by pulling the air molecules closer together. The further above the Earth's surface, the further apart the air molecules, and so the lower the density of the air. As the density of the air reduces with height, so does the pressure. At the top of Mount Everest (8848 m above sea level) the atmospheric pressure is about 30 per cent of atmospheric pressure at sea level.

Test yourself

34 Describe how the molecules in the atmosphere exert a pressure on a surface.
35 Why is the atmospheric pressure less at the top of a mountain than it is at the bottom of the mountain?
36 Four suction cups are used to lift a large piece of glass. The area of each cup is 0.04 m². Atmospheric pressure is 100 000 Pa. Calculate the weight of the largest piece of glass that can be lifted.

Show you can...

Show you understand the connection between the pull of gravity and atmospheric pressure by explaining why planets have different atmospheric pressures.

Chapter review questions

1 Which of these are scalar quantities and which are vector quantities?

   force   speed   mass   weight   temperature

2 Explain what is meant by the principle of moments.

3 The photograph shows sailors leaning out of their boat. Explain how this helps to balance the boat and stop it capsizing.

▲ Figure 5.43

4 Calculate the pull of the spring on the beam shown in Figure 5.44.

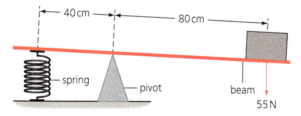

▲ Figure 5.44

5 a) i) Gravity is an example of a non-contact force. What does the term non-contact force mean?

   ii) Give an example of another non-contact force.

  b) A man has a mass of 85 kg. Calculate his weight. (g = 9.8 N/kg)

6 A boy with a weight of 600 N climbs up the Eiffel Tower. The Eiffel Tower has a height of 300 m. Calculate the work the boy does against gravity.

7 A weight lifter trains by holding two 5 kg masses at arms' length for 30 s. How much work has been done in this time? Give a reason for your answer.

8 a) Explain what is meant by elastic deformation.

  b) A spring is stretched elastically. How could you show that energy is stored in the spring?

9 A force of 10 N extends a spring by 2.5 cm. Calculate the spring constant of the spring in N/m.

10 Figure 5.45 shows how Ted used a simple seesaw to weigh his holiday luggage. Calculate the weight of his luggage.

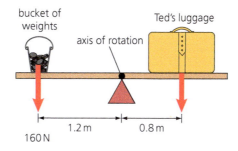

▲ **Figure 5.45**

11 a) A boy and a girl sit on a seesaw. It is balanced. Calculate the boy's weight.

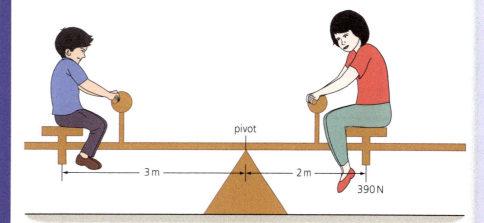

▲ **Figure 5.46**

b) The seesaw has a weight of 300 N. What force does the pivot exert on the seesaw?

 12 Use your knowledge of pressure to explain why the pressure increases as you dive towards the bottom of a swimming pool.

Practice questions

1 Which one of these quantities is a vector quantity?

| energy | time | velocity | [1 mark]

Give a reason for your choice. [1 mark]

2 A suitcase has a mass of 18 kg.

a) Calculate the weight of the case. [2 marks]

b) The case is lifted through a height of 2.1 m onto a luggage rack.

Calculate the work done in lifting it onto the luggage rack. [3 marks]

3 Figure 5.47 shows a machine for lifting water out of a lake.

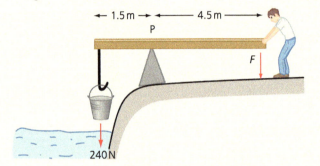

▲ Figure 5.47

a) Explain why the man using the machine pushes at the end of the lever. [2 marks]

b) Calculate the turning moment of the weight of the bucket about P. [2 marks]

c) Calculate the force exerted by the man to just balance the bucket of water. [3 marks]

4 Figure 5.48 shows a windsurfer in action.

▲ Figure 5.48

a) Explain why the windsurfer leans out on her sailboard. [2 marks]

b) The weight of the windsurfer is 750 N. Calculate the turning moment caused by this weight about the mast. [3 marks]

c) Explain why the windsurfer leans out further when the wind blows with greater force. [2 marks]

5 A passenger in an airport has a case with wheels.

▲ Figure 5.49

She decides it will be easier to pull the case if she packs all the heavy items of luggage nearest the wheels.

Explain why this is a good idea. Include ideas about turning moments in your answer. [4 marks]

6 a) State what is meant by pressure. [1 mark]

b) Explain briefly, in terms of pressure, why:

i) a tin can collapses when the air is removed from it [2 marks]

ii) a barometer (a device which measures atmospheric pressure) may be used to measure the altitude of an aircraft. [2 marks]

c) Figure 5.50 shows a simple type of hydraulic braking system. The areas of cross-section of the small cylinder and large cylinder are 0.0004 m² and 0.0024 m², respectively. A pedal is pushed against the piston in the small cylinder with a force of 90 N.

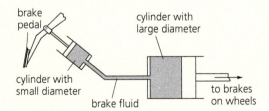

▲ Figure 5.50

i) Calculate the pressure exerted on the brake fluid. [3 marks]

ii) Calculate the force exerted by the brake fluid on the piston in the large cylinder. [2 marks]

7 Figure 5.51 shows a device that can be used to measure both small and large weights accurately. When a weight of 8 N is applied, the top plate rests on the lower fixed plates, so that only the lower three springs can stretch further.

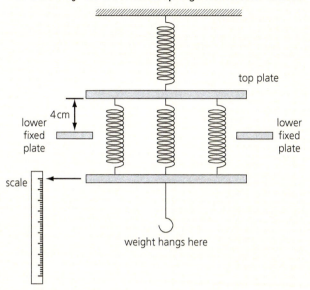

▲ Figure 5.51

The table shows the total extension of the springs as the weight is increased.

weight in N	Extension in cm
2.0	1.0
4.0	2.0
6.0	3.0
8.0	4.0
12.0	4.5
16.0	5.0
20.0	5.5
24.0	6.0
28.0	6.3
32.0	7.0

a) Plot a graph to show the weight against the extension of the springs. Draw a line of best fit through the points you have plotted. [4 marks]

b) The physicist doing the experiment has made a mistake with one of the measurements. Suggest what the measurement should have been. [1 mark]

c) Two weights are put on the device. One stretches the springs by 1.5 cm, the second stretches them by 5.2 cm. Use your graph to calculate each weight. [2 marks]

d) A student says: 'The graph shows that the extension of the springs is proportional to the weight.' Comment on this observation. [2 marks]

8 Figure 5.52 shows a mass of 24 kg which is placed on top of an electronic newton balance. A light string is attached to the mass and pulled upwards with a force of 100 N.

a i) Calculate the weight of the mass. [2 marks]

ii) Explain why the reading on the newton balance is 135 N. [2 marks]

The end of the string is now moved so that it pulls the mass a the angle shown in Figure 5.52 (b)

b i) Explain why the reading on the balance increases [1 mark]

ii) Show that the vertical component of the force exerted by the string on the mass is 87 N. [2 marks]

iii) Use a scale drawing to calculate the horizontal component of the force exerted by the string on the mass. [3 marks]

iv) The mass remains at rest on the newton balance. Explain why. [2 marks]

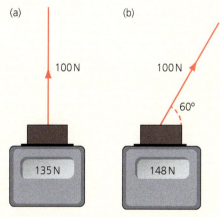

▲ Figure 5.52

Working scientifically

Hypothesis, prediction and testing

About 400 years ago, the scientist Galileo came up with the idea that all objects falling through the air would accelerate at the same rate. However, most people did not believe Galileo. They had seen with their own eyes that different-sized objects took different times to fall the same distance.

To show that his idea was right, Galileo did an experiment. He dropped a small iron ball and a large iron ball from the top of the Leaning Tower of Pisa. The two objects hit the ground at almost, but not quite, the same time.

1 What was the observation that made people say that Galileo's idea was wrong?

2 Why would the result of Galileo's experiment, on its own, not prove his idea?

Liam did an investigation to test Galileo's idea. Before starting, Liam wrote this hypothesis:

> 'The acceleration of a falling object is independent of the mass of the object.'

Using his hypothesis Liam then made a prediction:

> 'Objects of different mass will take the same time to fall the same distance.'

KEY TERMS

A **hypothesis** is an idea based on scientific theory. A hypothesis states what is expected to happen.

A hypothesis is used to make a **prediction** that can be tested. Experimental results that agree with the prediction provide evidence to support the hypothesis.

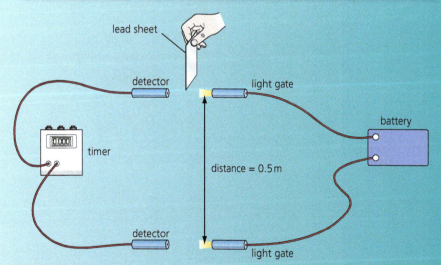

▲ **Figure 5.53** The apparatus used by Liam to test his hypothesis.

Liam dropped a small sheet of lead. As soon as he let go, the sheet passed through the top light gate and the electronic timer started. The timer stopped when the sheet passed through the bottom light gate. Liam repeated this three times. The three timer readings are shown in the table.

Result	1	2	3
Time in s	0.326	0.330	0.319

3 Why did Liam repeat the time measurements?

4 Liam used the three time values to calculate a mean. What was the mean time calculated by Liam?

5 Were the time values recorded by Liam **repeatable**? Give a reason for your answer.

Liam then timed lead sheets of the same area but increasing mass falling the same distance.

The mean time for each sheet is given in the table.

Sheet	Mass	Mean time in seconds
A	Smallest	0.320
B	Medium	0.325
C	Largest	0.322

6 Explain why the results from Liam's investigation support Galileo's idea about falling objects.

KEY TERM

Measurements are **repeatable** when the same person repeating the investigation under the same conditions obtains similar results.

5B Observing and recording motion

A catamaran such as this is capable of speeds of up to 50 mph (22 m/s). The catamaran's design is a feat of excellent engineering, which is based on many of the principles that you will meet in this chapter. The boat must be streamlined, and made of low density, high strength materials. It must be capable of high acceleration and rapid changes in direction.

Specification coverage

This chapter covers specification point: 4.5.6 Forces and motion. It covers 4.5.6.1 Describing motion along a line, 4.5.6.2 Forces, accelerations and Newton's Laws of Motion, 4.5.6.3 Forces and braking and 4.5.7 Momentum.

Prior knowledge

Previously you could have learned:

› A force can cause an object to speed up, slow down or change the direction of a moving object.
› When the resultant force acting on an object is zero, the object remains stationary or it moves at a constant speed in a straight line.
› Speed = $\dfrac{\text{distance travelled}}{\text{time}}$

Test yourself on prior knowledge

1 A parachutist falls from the sky at a steady speed. Are the forces on him balanced? Explain your answer.
2 a) A car accelerates along a road. In which direction is the resultant force on the car?
  b) Eventually the car reaches a constant speed. What is the resultant force now?
  c) The driver takes her foot off the accelerator. Explain why the car slows down.
3 a) A boy goes for a walk and covers a distance of 10 km in 3 hours 20 minutes. Calculate his average speed.
  b) A woman plans a car journey. She will cover a distance of 560 km at an average speed of 80 km per hour. She wants to include two breaks in the journey of 30 minutes. How long do you expect the journey to take?

Forces and motion

○ Describing motion along a line

The table shows some typical speeds of certain objects and people in motion. Some of these objects move at different speeds according to circumstances. A jet plane cruises at speeds of about 200 m/s, but takes off and lands at about 60 m/s.

Some typical speeds.

Moving object	Speed in m/s
Human walking	1.5
Human running	3.0
Human cycling	6.0
Car on the motorway	30
Express train	60
Jet plane	200

Human running speeds

A speed of 3 m/s represents the running speed of an average jogger we might see taking exercise in the park, but running speeds vary widely.

● We can run faster over a short distance. A 100 m sprinter runs much faster than a marathon runner.
● Young people run fast; we slow down as we get older.
● You cannot run so fast over rough terrain or when you are going uphill.

▲ **Figure 5.54** Hannah, Jenny and Natalia finish the 1500 m in 4 minutes and 5 seconds. What was their average speed?

Some people are extremely fit and strong. Calculate the speed of the athletes in Figure 5.54.

Average speed

When you travel in a fast car you finish a journey in a shorter time than when you travel in a slower car. If the speed of a car is 100 kilometres per hour (100 km/h), it will travel a distance of 100 kilometres in one hour.

Often, however, we use the equation to calculate an average speed because the speed of the car changes during the journey. When you travel along a motorway, your speed does not remain exactly the same. You slow down when you get stuck behind a lorry and speed up when you pull out to overtake a car.

The distance travelled by an object, moving at a constant speed, can be calculated using the equation:

$$s = v \times t$$

distance travelled = speed × time

where distance travelled is in metres, m

      speed is in metres per second, m/s

      time is in seconds, s.

Example

A train travels 440 km in 3 hours. What is its average speed in m/s?

Answer

To solve this problem, you need to remember that 1 km = 1000 m and that 1 hour = 3600 s.

$$\text{average speed} = \frac{s}{t}$$

$$= \frac{440 \times 1000}{3 \times 3600}$$

$$= 41 \, \text{m/s}$$

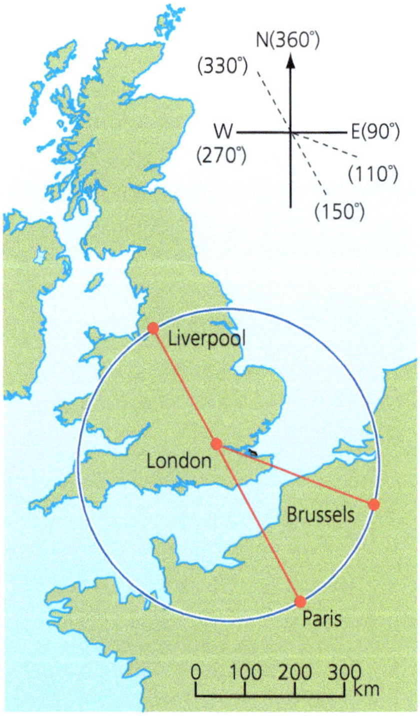

Vectors and scalars

The map in Figure 5.55 shows that Brussels, Paris and Liverpool are all a distance of about 300 km away from London. However, each city is in a different direction.

The displacement of Brussels from London is 300 km on a compass bearing of 110°. A helicopter can fly at about 300 km/h. So a helicopter can take us from London to one of the cities in an hour.

When a helicopter flies from London to Liverpool in an hour, its velocity is 300 km/h on a compass bearing of 330°.

Speed and distance are scalar quantities. Displacement and velocity are vector quantities because they are described by a magnitude and a direction.

▲ **Figure 5.55** Brussels, Paris and Liverpool are all a distance of about 300 km away from London.

Changing velocity

If you slow down when you are running, both your speed and velocity change. However, you can also change your velocity even though your speed remains constant. You will have done this while playing a game. When you play a game of football or netball, you may suddenly change your direction when running, but keep running at the same speed. Because a velocity is described by direction as well as speed, your velocity has changed.

When an object moves around a circular path, the velocity of the object changes all the time, even though the speed is constant. The direction of the motion changes, so the velocity changes too.

Here are some more examples of changing velocity while travelling in a circular path.

- When a car travels round a bend at constant speed its velocity changes.
- When an aeroplane banks to change direction its velocity changes.
- A satellite travelling round the Earth in a circular orbit travels at a constant speed. However, its direction is always changing. This is illustrated in Figure 5.56.

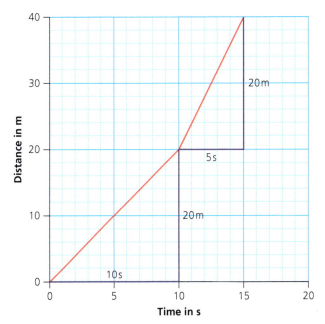

▲ Figure 5.56

> **TIP**
>
> An object that is travelling at a constant speed but has a changing velocity must be changing direction.

Distance–time graphs

When an object moves along a straight line, we can represent how far it has moved using a distance–time graph. Figure 5.57 is a distance–time graph for a runner. He sets off slowly and travels 20 m in the first 10 seconds. He then speeds up and travels the next 20 m in 5 seconds.

We can calculate the speed of the runner using the gradient of the graph.

▲ Figure 5.57

> **Example**
>
> Calculate the speed of the runner using the distance–time graph (Figure 5.57) a) over the first 10 seconds, b) over the time interval 10 s to 15 s.
>
> **Answer**
>
> a) $\text{speed} = \dfrac{\text{distance}}{\text{time}}$
>
> $\qquad = \dfrac{20\,\text{m}}{10\,\text{s}} = 2\,\text{m/s}$

b) $\text{speed} = \dfrac{40 - 20}{5}$

$= 4\,\text{m/s}$

You can see from the graph that these speeds are the gradients of each part of the graph.

H When you set off on a bicycle ride, it takes time for you to reach your top speed. You accelerate gradually.

Figure 5.58 shows a distance–time graph for a cyclist at the start of a ride. The gradient gets steeper as time increases. This tells us that his speed is increasing. So the cyclist is accelerating.

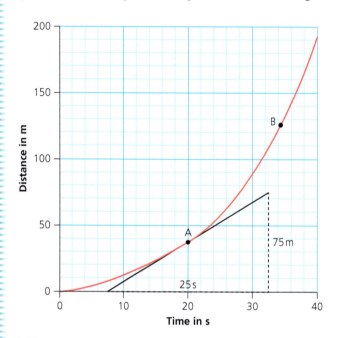

▲ **Figure 5.58**

We can calculate the speed of the cyclist at any point by drawing a tangent to the curve, and then measuring the gradient.

Example

On the graph in Figure 5.58, a gradient has been drawn at point A, 20 seconds after the start of the ride. What is the speed at this time?

Answer

$\text{speed} = \text{gradient} = \dfrac{\text{vertical height (distance)}}{\text{horizontal length (time)}}$

$\text{speed} = \dfrac{75}{25}$

$= 3\,\text{m/s}$

Test yourself

1 A helicopter flies from London to Paris in 2 hours. Use the information in Figure 5.55 to calculate
   a) the helicopter's speed
   b) the helicopter's velocity.
2 A car travels 100 m at a speed of 20 m/s. Sketch a distance–time graph to show the motion of the car.
3 Curtis cycles to school. Figure 5.59 shows the distance–time graph for his journey.
   a) How long did Curtis stop at the traffic lights?
   b) During which part of the journey was Curtis travelling fastest?
4 The table below shows average speeds and times recorded by top male athletes in several track events. Copy and complete the table.

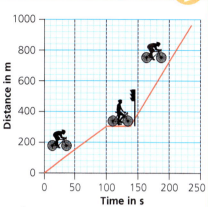

▲ Figure 5.59

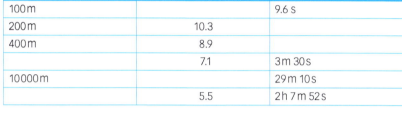
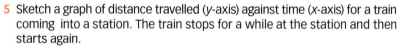

Event	Average speed in m/s	Time
100 m		9.6 s
200 m	10.3	
400 m	8.9	
	7.1	3 m 30 s
10 000 m		29 m 10 s
	5.5	2 h 7 m 52 s

5 Sketch a graph of distance travelled (*y*-axis) against time (*x*-axis) for a train coming into a station. The train stops for a while at the station and then starts again.
6 Which of the answers below is the closest to the speed at point B of Figure 5.58?
   4 m/s   10 m/s   15 m/s
7 Explain why it is possible to travel at a constant speed, but have a changing velocity.
8 Ravi, Paul and Tina enter a 30 km road race. Figure 5.60 shows Ravi's and Paul's progress through the race.
   a) Which runner ran at a constant speed? Explain your answer.
   b) What was Paul's average speed for the 30 km run?
   c) What happened to Paul's speed after 2 hours?

Tina was one hour late starting the race. During the race she ran at a constant speed of 15 km/h.
   d) Copy the graph and add to it a line to show how Tina ran.
   e) How far had Tina run when she overtook Paul?

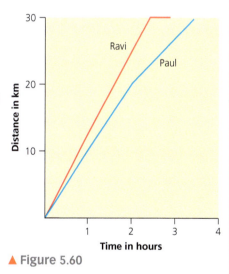

▲ Figure 5.60

Show you can...

Show that you understand distance–time graphs by completing this task.

Draw a distance–time graph to show this motion. A man walks for 30 seconds at 2 m/s; he stops for 20 s; he then runs at 6 m/s for 10 s.

Include as much information as possible in your graph. Explain what the gradient of the graph shows at each point.

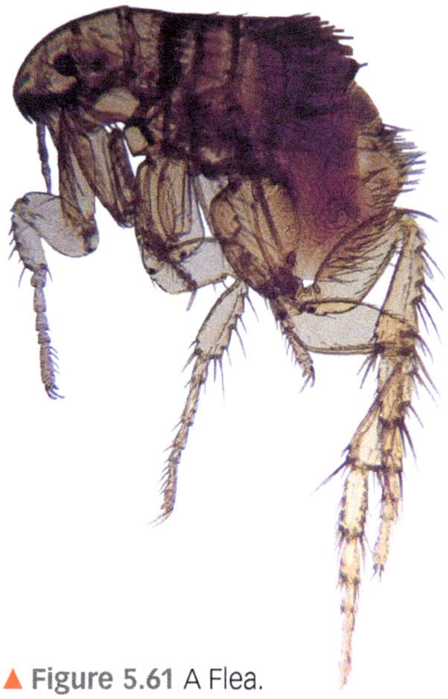

◯ Acceleration

Starting from the grid, a Formula 1 car reaches a speed of 30 m/s after 2 s. A flea can reach a speed of 1 m/s after 0.001 s. Which accelerates faster?

Speeding up and slowing down

When a car is speeding up, we say it is accelerating. When it is slowing down we say it is decelerating.

A car that accelerates rapidly reaches a high speed in a short time. For example, a car might speed up to 12.5 m/s in 5 seconds. A van could take twice as long, 10 s, to reach the same speed. So the acceleration of the car is twice as big as the van's acceleration.

You can calculate the acceleration of an object using the equation:

$$a = \frac{\Delta v}{t}$$

$$\text{acceleration} = \frac{\text{change in velocity}}{\text{time taken}}$$

where acceleration is in metres per second squared, m/s^2

change in velocity (Δv) is in metres per second, m/s

time taken is in seconds, s

$$\text{acceleration of car} = \frac{12.5}{5} = 2.5\,m/s^2$$

$$\text{acceleration of van} = \frac{12.5}{10} = 1.25\,m/s^2$$

▲ Figure 5.61 A Flea.

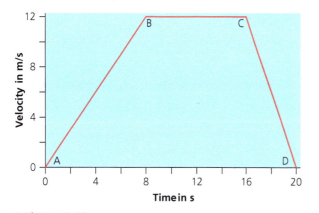

▲ Figure 5.62 A Formula 1 racing car.

Velocity–time graphs

It can be helpful to plot graphs of velocity against time.

Figure 5.63 shows the velocity–time graph for a cyclist as she goes on a short journey along a straight road.

The area under a velocity–time graph represents the distance travelled (or displacement).

- In the first 8 seconds, the cyclist accelerates up to a speed of 12 m/s (section AB of the graph).
- For the next 8 seconds, she cycles at a constant speed (section BC of the graph).
- Then for the last 4 seconds of the journey, she decelerates to a stop (section CD of the graph).

The gradient of the graph gives us the acceleration. In section AB she increases her speed by 12 m/s in 8 seconds.

$$\text{acceleration} = \frac{\text{change in velocity}}{\text{time}}$$
$$= \frac{12}{8} = 1.5\,m/s^2$$

▲ Figure 5.63

TIP

The gradient of a velocity–time graph represents the acceleration.
The area under section AB gives the distance travelled because:

distance = average speed × time

= 6 m/s × 8 s

= 48 m

The average speed is 6 m/s as this is half of the final speed 12 m/s. The area can also be calculated using the formula for the area of the triangle:

area of triangle = $\frac{1}{2}$ × base × height

= $\frac{1}{2}$ × 8 s × 12 m/s

= 48 m

The following equation applies to an object that has a constant acceleration.

$$v^2 - u^2 = 2\,a\,s$$

where v is the final velocity in metres per second, m/s

u is the initial (starting) velocity in metres per second, m/s

a is the acceleration in metres per second squared, m/s^2

s is the distance in metres, m.

Provided the value of three of the quantities is known, the equation can be rearranged to calculate the unknown quantity.

Example

The second stage of a rocket accelerates at 3 m/s^2. This causes the velocity of the rocket to increase from 450 m/s to 750 m/s. Calculate the distance the rocket travels while it is accelerating.

Answer

$v = 750$ m/s, $u = 450$ m/s, $a = 3$ m/s^2

$v^2 - u^2 = 2\,a\,s$

$750^2 - 450^2 = 2 × 3 × s$

$562\,500 - 202\,500 = 6 × s$

$s = \dfrac{360\,000}{6} = 60\,000$ m

When the acceleration is not constant, you can work out the distance travelled by working out the area under a velocity-time graph. We do this by counting the squares, but we need to work out what distance each square represents.

Example

A hot air balloon rises from the ground. Figure 5.64 is a velocity–time graph which shows how the balloon's velocity changes with time. Use the graph to calculate the distance the balloon rises, before stopping.

Answer

We count the squares under the graph. There are 20 whole squares. Then we have to estimate the area of the incomplete squares – this is about 7 (check that you agree.)

So the total area is 27 squares.

Each square represents a distance of 2 m/s × 4 s = 8 m.

Therefore the distance travelled is 27 × 8 m = 216 m.

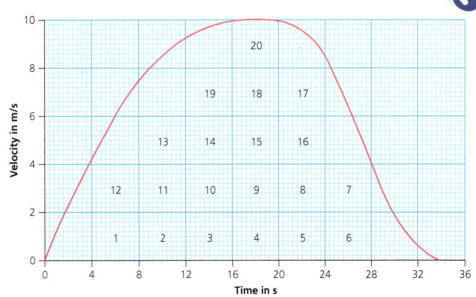

▲ **Figure 5.64**

Show you can...

Show that you understand that velocity–time graphs are useful by completing this task.

A car accelerates from rest to 10 m/s over 20 s. It then travels at a constant speed for 15 s. Sketch a velocity–time graph to show this motion. Explain how the graph may be used to calculate the car's acceleration and the distance the car travels.

Test yourself

9 Which is the correct unit for acceleration?
   ms^2 ms   m/s m/s^2

10 a) Write down the equation which links acceleration to a change of velocity and time.
   b) Use the information at the top of page 152 to calculate the acceleration of:
      i) the F1 car
      ii) the flea.

11 This question refers to the journey shown in Figure 5.63.
   a) Calculate the cyclist's deceleration over region CD of the graph.
   b) Use the area under the graph to calculate the distance covered during the whole journey.
   c) Calculate the average speed over the whole journey.

12 The table shows how the speed, in m/s, of a Formula 1 racing car changes as it accelerates away from the starting grid at the beginning of a Grand Prix.
   a) Plot a graph of speed (y-axis) against time (x-axis).
   b) Use your graph to calculate the acceleration of the car at:
      i) 16 s
      ii) 1 s.
   c) Calculate how far the car has travelled after 16 s.

Speed in m/s	0	10	20	36	49	57	64	69	72	72
Time in s	0	1	2	4	6	8	10	12	14	16

13 Copy the table and fill in the missing values.

	Starting speed in m/s	Final speed in m/s	Time taken in s	Acceleration in m/s^2
Cheetah	0		5	6
Train	13	25	120	
Aircraft taking off	0		30	2
Car crash	30	0		−150

14 Drag cars are designed to cover distances of 400 m in about 6 seconds. During this time the cars accelerate very rapidly from a standing start. At the end of 6 seconds, a drag car reaches a speed of 150 m/s.
   a) Calculate the drag car's average speed.
   b) Calculate its average acceleration.

15 An aeroplane that is about to take off accelerates along the runway at 2.5 m/s^2. It takes off at a speed of 60 m/s. Calculate the minimum length of runway needed at the airport.

○ Observing motion

A trainer who wants to know how well one of his athletes is running stands at the side of the track with a stopwatch in his hand. By careful timing, the athlete's speed at each part of the race can be analysed. However, if the trainer wants to know more detail, the measurements need to be taken with smaller intervals of time. One way to do this is by filming the athlete's movements. The following practicals show two further ways of analysing motion.

▲ Figure 5.65 This high-speed photograph allows the athlete to improve his technique by analysing his motion.

Practical

Light gates

The speed of a moving object can also be measured using light gates. Figure 5.66 shows an experiment to determine the acceleration of a rolling ball as it passes between two light gates. When the ball passes through a light gate, it cuts a beam of light. This allows the computer to measure the time taken by the ball to pass through the gate. By knowing the diameter of the ball, the speed of the ball at each gate can be calculated. You can tell if the ball is accelerating if it speeds up between light gates A and B. Follow the questions below to show how the gates can be programmed to do the work for you.

The measurements taken in an experiment are shown here.

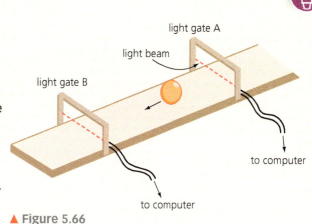

▲ Figure 5.66

Diameter of the ball	6.2 cm
Time for the ball to go through gate A	0.12 s
Time for the ball to go through gate B	0.09 s
Time taken for the ball to travel from gate A to gate B	0.23 s

1 Explain why it is important to adjust the light gates to the correct height.
2 Calculate the speed of the ball as it goes through:
  a) gate A
  b) gate B.
3 Calculate the ball's acceleration as it moves from gate A to gate B.

Ticker timer

Changing speeds and accelerations of objects in the laboratory can be measured directly using light gates, data loggers and computers. However, motion is still studied using the ticker timer (Figure 5.67), because it collects data in a clear way, which can be usefully analysed. A ticker timer has a small hammer that vibrates up and down 50 times per second. The hammer hits a piece of carbon paper, which leaves a mark on a length of tape.

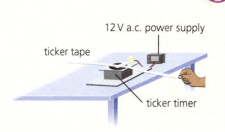

▲ Figure 5.67

Figure 5.68 shows you a tape that has been pulled through the timer. You can see that the dots are close together over the region PQ. Then the dots get further apart, so the object moved faster over QR. The movement slowed down again over the last part of the tape, RS. Since the timer produces 50 dots per second, the time between dots is 1/50 s or 0.02 s. So we can work out the speed:

$$\text{speed} = \frac{\text{distance between dots}}{\text{time between dots}}$$

Between P and Q, speed = $\dfrac{0.5\,\text{cm}}{0.02\,\text{s}}$

= 25 cm/s or 0.25 m/s

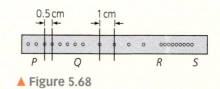

▲ Figure 5.68

Questions

1 Work out the speed of the tape in the region QR.
2 Which is the closest to the speed in the region RS?
  0.1 m/s   0.3 m/s   0.8 m/s

▲ **Figure 5.69** The skydiver has reached terminal velocity in a streamlined position. How could the skydiver slow down before opening the parachute?

○ Falling objects and parachuting

A falling object accelerates due to the pull of gravity. If there were no air resistance, an object dropped from any height would accelerate at 9.8 m/s² until it hit the ground. You can demonstrate that two objects of different mass accelerate at the same rate over a short distance. A marble and a 100 g mass dropped from a height of 2 m reach the ground at the same time. However, if you drop a feather it flutters towards the ground because the air resistance on it is about the same size as its weight.

Parachuting

The size of the air resistance on an object depends on the area of the object and its speed.

- The larger the area, the larger the air resistance.
- The larger the speed, the larger the air resistance.

Figure 5.70 shows how the velocity of a skydiver changes as she falls directly to the ground. The graph has five distinct parts.

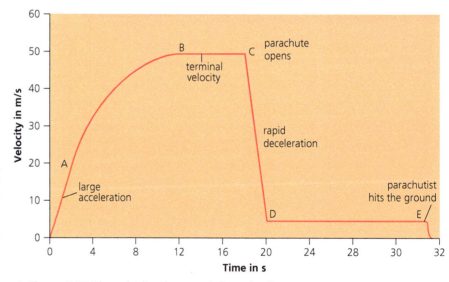

▲ **Figure 5.70** The velocity–time graph for a skydiver.

TIP

It does not matter what the object is, if it has reached its terminal velocity then the resultant force must be zero.

1 OA. She accelerates at about 9.8 m/s² just after leaving the aeroplane.

2 AB. The effects of air resistance make her acceleration less and the gradient of the graph gets less. But she continues to accelerate as her weight is greater than the air resistance. There is a resultant force downwards.

KEY TERM

An object falls at its **terminal velocity** when its weight is balanced by resistive forces.

3 BC. The air resistance has now increased so that it is the same size as her weight. The resultant force is zero, and she moves at a constant velocity. This is her **terminal velocity**.

4 CD. She opens her parachute. The increased surface area causes an increased air resistance. So she decelerates and slows down as there is a resultant force upwords.

5 DE. The air resistance on the parachute is now the same as her weight. She continues at a slower terminal velocity until she hits the ground.

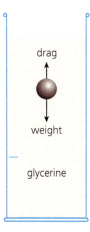

▲ **Figure 5.71** A ball bearing falling through a thick liquid (like glycerine) reaches its terminal velocity before an identical ball bearing falling through the air reaches a faster velocity. Which is greater, the drag force in the glycerine or air resistance, when the two balls fall at their terminal velocities?

▲ **Figure 5.72** The top speed of a car is affected by the power of the engine and air resistance. By making the car streamlined, air resistance reduces and the top speed increases.

Practical

Terminal velocity and surface area

Figure 5.73 shows a model parachute made from a sheet of plastic.

Plan an experiment to find out whether the terminal velocity of the Plasticine figure depends on the area of the parachute.

Questions

1 Include a risk assessment in your plan.
2 What measurements will you make?
3 How will you know that the Plasticine figure has reached its terminal velocity?
4 Which one of the following is the independent variable in this experiment?

distance the parachute falls

area of the parachute

weight of the Plasticine figure

time taken to fall.

5 What are the control variables in this experiment?

▲ **Figure 5.73**

Test yourself

16 A parachutist falls at a constant speed. Which of the following statements is correct? Copy out the correct statement.
   • Her weight is much more than the air resistance.
   • Her weight is just a little more than the air resistance.
   • Her weight is the same size as the air resistance.
17 What is meant by terminal velocity?
18 Explain this observation: 'When a sheet of paper is dropped it flutters down to the ground, but when the same sheet of paper is screwed up into a ball, it accelerates rapidly downwards when dropped.'
19 This question refers to the velocity–time graph in Figure 5.70.
   a) What was the velocity of the skydiver when she hit the ground?
   b) Why is her acceleration over the part AB less than it was at the beginning of her fall?
   c) Use the graph to estimate roughly how far she fell during her dive. Was it nearer 100 m, 1000 m, or 10 000 m?
   d) Draw diagrams to show the size of the skydiver's weight (*W*), and the air resistance (*D*) acting on her between these points:
      i) OA   ii) BC   iii) CD.

Forces, accelerations and Newton's Laws of Motion

◯ Newton's first law: balanced forces

When the resultant force acting on an object is zero, the forces are balanced and the object does not accelerate. It remains stationary, or continues to move in a straight line at a constant speed.

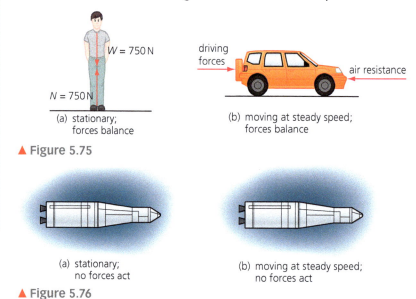

(a) stationary;
    forces balance

(b) moving at steady speed;
    forces balance

▲ Figure 5.75

(a) stationary;
    no forces act

(b) moving at steady speed;
    no forces act

▲ Figure 5.76

Figure 5.75 and 5.76 show some examples where the resultant force is zero.

- A person is standing still. Two forces act on him: his weight downwards and the normal contact force from the floor upwards. The forces balance; he remains stationary.
- A car moves along the road. The forwards push from the road on the car is balanced by the air resistance on the car. The forces are balanced and so the car moves with a constant speed in a straight line.
- A spacecraft is in outer space, so far away from any star that the gravitational force is zero. There are no frictional forces. So the resultant force is zero. The spacecraft is either at rest or moving in a straight line at a constant speed.

The speed and/or direction of an object will only change if a resultant force acts on the object. Newton's first law does not apply to these objects.

◯ Newton's second law: unbalanced forces

When an unbalanced force acts on an object it accelerates. The object could speed up, slow down or change direction.

Figure 5.77 shows two examples of unbalanced forces acting on a body.

a) The spacecraft has turned its rockets on. There is a force pushing the craft forwards, so it accelerates.

b) The driver has taken his foot off the accelerator while the car is moving forwards. There is an air resistance force that acts to decelerate the car.

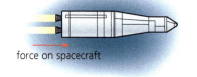

(a) acceleration

(b) deceleration

▲ Figure 5.77

▲ Figure 5.74 The Saturn V rocket is the largest rocket ever to take off. It took men to the Moon in 1969. Planning its flight required an understanding of Newton's laws.

Force, mass and acceleration

You may have seen people pushing a car with a flat battery. When one person tries to push a car, the acceleration is very slow. When three people give the car a push, it accelerates more quickly.

You will know from experience that large objects are difficult to get moving. When you throw a ball you can accelerate your arm more quickly if the ball has a small mass. You can throw a tennis ball much faster than you put a shot. A shot has a mass of about 7 kg so your arm cannot apply a force large enough to accelerate it as rapidly as a tennis ball.

Newton's second law states that

- acceleration is proportional to the resultant force

$$a \propto F$$

- acceleration is inversely proportional to the mass

$$a \propto \frac{1}{m}$$

This can be written as an equation:

$$F = ma$$
resultant force = mass × acceleration

where force is in newtons, N

mass is in kilograms, kg

acceleration is in metres per second squared, m/s².

TIP

In the equation $F = ma$, F means resultant force.

Example

1000 N   640 N

▲ **Figure 5.78**

The mass of the car in Figure 5.78 is 1200 kg. Calculate the acceleration.

Answer

$$F = ma$$
$$1000 - 640 = 1200 \times a$$
$$a = \frac{360}{1200}$$
$$= 0.3 \, \text{m/s}^2$$

Note

- First we had to work out the resultant force.
- The acceleration is in the same direction as the resultant force. So the car speeds up or accelerates.

TIP

When you use the equation $F = ma$, the force must be in N, the mass in kg and the acceleration in m/s².

Investigating the relationship between force and acceleration

Figure 5.79 shows one way that you can investigate the acceleration of an object.

The accelerating force is caused by the falling mass. A falling mass of 100 g tied to the string exerts a force of 1 N.

A light gate is being used in this investigation but you could use a ticker timer and tape.

Use the method given below to investigate the following hypothesis:

the acceleration of an object is directly proportional to the force applied to the object.

Using this hypothesis we can predict that:

doubling the force applied to the object will double the acceleration of the object.

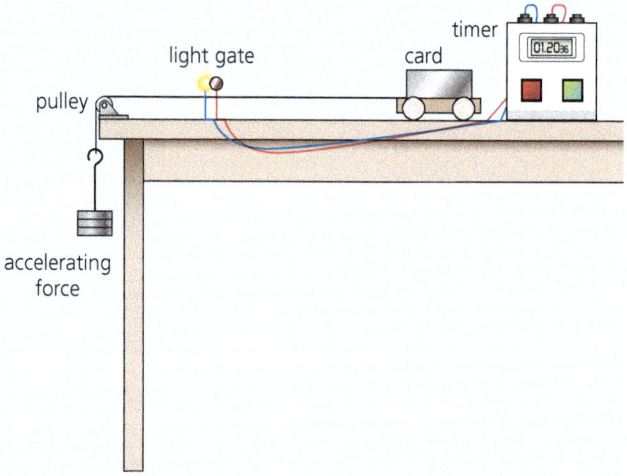

▲ Figure 5.79

Method

1 Set up the equipment as shown in Figure 5.79.

2 The timer should start and then stop as the card passes through the light gate.

3 Draw a table to record all of the data collected. Include a column for the acceleration of the trolley.

4 Tie a mass of 100 g to the string.

5 Add two extra 100 g masses to the trolley.

6 Place the trolley so that the front of the trolley is 0.5 m away from the light gate. The string should be taut and the hanging mass at least 0.5 m above the floor. Hold the trolley stationary so its initial velocity is zero.

7 Let go of the trolley and use a stopwatch to time how long the trolley takes to travel to the light gate.

   Make sure that you are able to stop the trolley from falling to the floor and keep your feet away from the falling masses.

8 Write down the time taken for the card to pass through the light gate.

9 By taking a 100 g mass from the top of the trolley and adding it to the hanging masses you increase the accelerating force. You can now gather data for a range of accelerating forces.

Calculating the acceleration

1 $$\text{final velocity of the trolley} = \frac{\text{length of card}}{\text{time taken to pass through the light gate}}$$

2 $$\text{acceleration} = \frac{\text{change of velocity}}{\text{time taken}} = \frac{\text{final velocity of the trolley}}{\text{time taken to reach the light gate}}$$

If you use a ticker timer, attach a 1 m length of ticker tape to the back of the trolley. Pass the tape through the timer. When you let go of the trolley a pattern of dots will be printed on the tape. Use this pattern to calculate the acceleration of the trolley.

Analysing the results

1 Plot a graph of acceleration against force.

2 Do the results from your investigation support the hypothesis? Explain the reason for your answer.

Taking it further

1 Plan and carry out an investigation to find out how the mass of an object affects the acceleration of the object.

2 Before starting your investigation, write your own hypothesis. Use your hypothesis to predict what will happen.

3 Write a **risk assessment** for your investigation.

Explain why it is necessary to store the extra 100 g masses on top of the trolley which are later used to increase the accelerating force.

Questions

1 Why are the velocity values you calculate average values?

2 The trolley could have been run along a slightly raised runway rather than along a level bench top. Explain the effect this would have on the values calculated for acceleration.

3 In the first investigation, mass is kept constant while the accelerating force is changed. If you did the second investigation, the accelerating force will have been kept constant while the mass is changed. Why is it important that either the mass or the accelerating force is kept constant?

KEY TERM

In a **risk assessment** you should identify:

• the hazards
• the possible risks associated with each hazard
• any control measures you can take to reduce the risks.

Show you can...

Show you understand the connection between force, mass and acceleration by completing this task.

Design a practical that enables you to show that an object accelerates at a greater rate when a greater force is applied to it. State clearly how you will measure the acceleration of the object.

Test yourself

20 Explain why Formula 1 racing cars have low masses.

21 Explain why you can throw a tennis ball faster than you can throw a 7 kg shot.

22 a) Calculate the resultant force acting on a mass of 8 kg which is accelerating at 2.5 m/s^2.

  b) Calculate the acceleration of a 3 kg mass that experiences a 15 N resultant force.

  c) Calculate the mass of an object which is accelerating at 4 m/s^2 when a resultant force of 10 N acts on it.

▲ **Figure 5.80** How do you know whether there are biscuits left without taking the lid off?

Inertia

Imagine that you are in a spacecraft on a long voyage to Mars. You are in the kitchen and there are two very large boxes of biscuits, marked with their mass – 14 kg. One box is empty, one box is full. How can you tell which is full without taking the lid off? Remember both boxes have no weight, because there is no gravitational pull on them.

The answer to the question above is that you give each of the boxes a small push. The empty one is easy to set in motion, but the one full of biscuits is much harder to start moving.

We say that the box full of biscuits has more **inertia** than the empty box.

KEY TERM

Inertia is a Latin word, which means inactivity. Here we use the word to mean that objects remain in their existing state of motion – at rest or moving with a constant speed in a straight line – unless acted on by an unbalanced force.

The inertial mass of an object is a measure of how difficult it is to change its velocity.

We can define the inertial mass through the equation:

$$m = \frac{F}{a}$$

In the case of the biscuit tins, the one with the smaller acceleration (for the same applied force) has the larger inertial mass.

Test yourself

23 Explain why oil tankers (which have masses of about half a million tons) take about 20 minutes to stop moving when their engines are turned off.

24 A Formula 1 car has a mass of 730 kg. The car brakes from a speed of 84 m/s to 32 m/s in 1.3 s.
   a) Calculate the deceleration during this time.
   b) Calculate the size of the braking force on the car.

25 A spacecraft on the Moon has a weight of 48 000 N. The mass of the spacecraft is 30 000 kg.
   a) Calculate the Moon's gravitational field strength.
   b) When the spacecraft takes off, the force on the spacecraft due to the rockets is 63 000 N.
      i) Calculate the resultant force on the spacecraft.
      ii) Calculate the acceleration as the spacecraft takes off.

26 Explain why a passenger, who is standing, might lose his balance if a train accelerates quickly out of a station.

27 A parcel is on the seat of a car. When the car brakes suddenly, the parcel moves forwards and falls off the seat and on to the floor. Explain why.

Show you can...

Show that you understand Newton's first law of motion by completing this task.

Imagine you have just become a physics teacher. Plan a lesson with three experiments that help to explain Newton's first law of motion.

○ Newton's third law of motion

Every force has a paired equal and opposite force. This law sounds easy to apply, but it requires some clear thinking. It is important to appreciate that the pairs of forces must act on *two* different objects, and the forces must be the same type of force.

An easy way to demonstrate Newton's third law is to connect two dynamics trolleys together with a stretched rubber band, as shown in Figure 5.81.

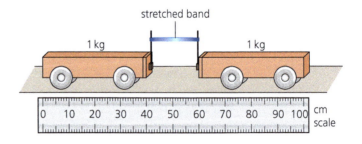

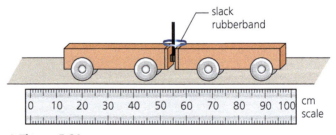

▲ Figure 5.81

When the trolleys are released, they travel the same distance and meet in the middle. This is because trolley A exerts a force on trolley B, and trolley B exerts a force of the same size, in the opposite direction, on trolley A. Some examples of paired forces are given in Figure 5.82.

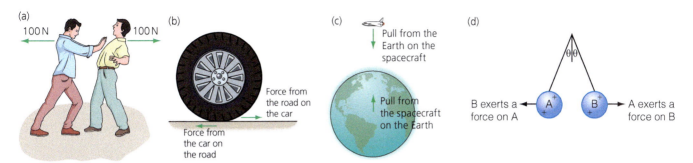

▲ **Figure 5.82** a) If I push you with a force of 100 N, you push me back with a force of 100 N. b) When the wheel of a car turns, it pushes the road backwards. The road pushes the wheel forwards with an equal and opposite force. c) A spacecraft orbiting the Earth is pulled downwards by the Earth's gravity. The spacecraft exerts an equal and opposite gravitational force on the Earth. So if the spacecraft moves towards the Earth, the Earth moves too, but because the Earth is so massive its movement is very small. d) Two balloons have been charged positively. They each experience a repulsive force from the other. These forces are of the same size, so each balloon (if of the same mass) is lifted through the same angle.

All of these are features of Newton's third law pairs:

- they act on two separate bodies
- they are always of the same type, for example two electrostatic forces or two contact forces
- they are of the same magnitude
- they act along the same line
- they act in opposite directions.

Show you can...

Show you understand Newton's third law of motion by explaining three demonstrations of the law to a friend.

Test yourself

28 Use Newton's third law to explain the following.
 a) When you lean against a wall you do not fall over.
 b) When you are swimming you have to push water backwards so that you can move forwards.
 c) You cannot walk easily on icy ground.
29 A man with a weight of 850 N jumps off a wall of a height 1.2 m. While he is falling, what force does he exert on the Earth?

Forces and braking

When you begin to drive the most important thing you must learn is how to stop safely.

Frictional forces decelerate a car and bring it to rest. However, the driver also needs to use the brakes to slow the car in traffic or when traffic lights turn red. In an emergency the driver must apply the brakes as soon as possible. However, the average driver has a reaction time of about 0.7 s. This is the time taken by the driver to react to an emergency, and move their feet from the accelerator pedal to the brake pedal.

KEY TERMS

The **thinking distance** is the distance a car travels while the driver reacts.

The **braking distance** is the distance a car travels while the car is stopped by the brakes.

The **stopping distance** is the sum of the thinking distance and braking distance.

During this time the car carries on moving at a constant speed. The distance the car moves during the reaction time is called the **thinking distance**.

Once the brakes have been applied, the car slows down and stops. The distance the car moves once brakes are applied is called the **braking distance**.

The **stopping distance** of a vehicle (Figure 5.83) is made up of the two parts – the thinking distance and the braking distance.

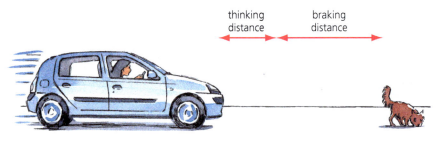

▲ Figure 5.83

Speed affects both the thinking distance and the braking distance. Figure 5.84 shows the stopping distances for a typical car at different speeds, in good conditions, on a dry road. The force applied by the brakes is the same for each speed.

▲ **Figure 5.84** These stopping distances appear in the Highway Code. You need to know these to drive safely. The reaction time here is about 0.7 s.

Factors affecting reaction times

Reaction times vary from person to person. Typical values range from about 0.2 s to 0.9 s.

Drivers' reaction times are much slower if they:

- have been drinking alcohol
- have been taking certain types of drugs
- are tired.

It is not just illegal drugs that slow reactions. Anybody taking a medicine and intending to drive should check the label.

People can also react slowly if they are distracted, either by thinking about something else or talking on their phone.

TIP

You should recall that a typical reaction time for a driver is 0.7s.

TIP

Factors that affect thinking distance are to do with the driver. Factors that affect braking distance are to do with the weather, the road or the car.

Factors affecting braking distances

Braking distances can be affected by several factors as well as speed:

- the size of the braking force – if you brake harder, a larger force acts to decelerate the car, so you stop in a shorter distance; but you have to be careful if the force is too large because a car can skid and you may lose control
- weather conditions – in wet or icy conditions there is less friction between the tyre and the road; this increases the braking distance
- the vehicle is poorly maintained – worn brakes or worn tyres increase the braking distance
- the road surface – a rough surface increases the friction between the road and tyres and reduces the braking distance.

Braking and kinetic energy

The work done by a braking force reduces a vehicle's kinetic energy, causing the vehicle to slow down and stop. As the kinetic energy of the vehicle goes down, the temperature of the brakes goes up. This is why friction brakes always get hot when they are applied. Large decelerations caused by large braking forces may lead to the brakes overheating.

See page 124 for more information about work done.

Practical

Reaction and distraction

Method

Hold a ruler between the fingers of a partner's hand. Without giving a warning, let go of the ruler. Your partner should close their fingers as soon as they see the ruler fall.

Try it again, but this time with your partner listening to music or talking to a friend. What do you predict will happen?

If there is a difference in reaction time, how could you check that it was due to the music and not some other factor?

Test yourself

30 **a)** What is meant by **i)** thinking distance, **ii)** braking distance, **iii)** stopping distance?

   b) What is the connection between thinking distance, braking distance and stopping distance?

31 Which one of the following would affect the braking distance of a car?
   a tired driver   an icy road   a drunk driver

32 Copy and complete the following sentences:

   a) For a particular braking _____ the greater the _____ of a car, the greater the braking distance.

   b) A driver's reaction time may be increased if the driver is using a _____ .

33 How could a road surface near a busy junction be changed to help reduce the braking distance of a car approaching the junction?

34 Worn tyres are dangerous. The graph in Figure 5.85 shows how the braking distance for a car depends on the depth of the tread on its tyres. Two road surfaces have been tested.

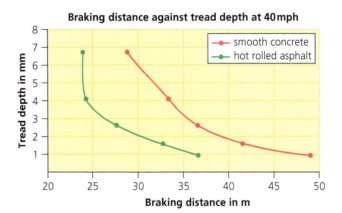

Braking distance against tread depth at 40 mph

▲ Figure 5.85

a) How do the curves show that worn tyres are dangerous?

b) Explain the importance of the road surface on braking distances.

35 a) Use Figure 5.84 to explain why it is so dangerous to drive at 40 mph in a 30 mph speed limit.

b) Explain why some councils impose 20 mph speed limits in some areas.

c) Use the data in Figure 5.84 to show that the driver's reaction time is about 0.7 s.

36 Use the data in Figure 5.84 to calculate the deceleration of a car coming to a stop when travelling initially at:

a) 20 mph

b) 60 mph.

[Hint: you might need to refer back to the equation on page 152.]

[Hint: you might need to refer back to the equation on page 152.]

Show you can...

Show you understand about stopping distances by completing this task.

Write a paragraph to explain the factors which affect the stopping distance for a car.

Momentum

○ Momentum is a property of moving objects

Momentum is defined as the product of mass and velocity

$$p = m \times v$$

momentum = mass × velocity

where momentum is in kilogram metres per second, kg m/s

mass is in kilograms, kg

velocity is in metres per second, m/s.

▲ Figure 5.86 Why does a shotgun recoil when it is fired?

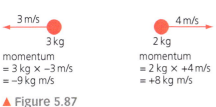

3 m/s
3 kg
momentum
= 3 kg × –3 m/s
= –9 kg m/s

4 m/s
2 kg
momentum
= 2 kg × +4 m/s
= +8 kg m/s

▲ **Figure 5.87**

Velocity is a vector quantity and therefore so is momentum. This means you must give a size and a direction when you talk about momentum. Look at Figure 5.87, where you can see two objects moving in opposite directions. One has positive momentum and the other negative momentum. Here we have chosen to define the positive direction to the right.

○ Changes in momentum

When a force, F, pushes a mass, m, we can relate it to the mass and acceleration, using the equation:

$$F = ma$$
$$\text{force} = \text{mass} \times \text{acceleration}$$

The acceleration is defined by the equation:

$$\text{acceleration} = \frac{\text{change of velocity}}{\text{time}}$$

These two equations can be combined to give:

$$\text{force} = \frac{\text{mass} \times \text{change of velocity}}{\text{time}}$$

Because (mass × change of velocity) is the change of momentum, the force can be written as:

$$\text{force} = \frac{\text{change of momentum}}{\text{time}}$$

So the force applied equals the rate of change of momentum. Rate of change means how fast the quantity (in this case momentum) the momentum changes with time.

Example

A mass of 7 kg is accelerated from a speed of 2 m/s to 6 m/s in 0.5 s. What force must be applied to the mass to do this?

Answer

$$\text{force} = \frac{\text{change of momentum}}{\text{time}}$$
$$= \frac{(7 \times 6) - (7 \times 2)}{0.5}$$
$$= 56 \, \text{N}$$

Momentum and safety

The equation relating force and change of momentum is important when it comes to considering a number of safety features in our lives. If you are moving, you have momentum. To stop moving, a force must be applied. The equation shows us that if we stop over a long period of time, the force to slow us is smaller. This way we get hurt less.

Here are some examples of how we protect ourselves by increasing the time we or another object has to stop.

- When we catch a ball that is moving fast, we move our hands backwards with the ball.
- We wear shin pads in hockey so that the ball has more time to stop when it hits us.
- When we jump off a wall we bend our legs when we land, so that we take a longer time to stop.
- Children's playgrounds have soft, rubberised matting under climbing frames, so that a falling child takes a longer time to stop.
- In a bungee jump, the elastic rope stretches to slow your fall gradually.

Driving safely

Figure 5.88 shows two cars being tested in a trial crash. In the centre of the cars are rigid passenger cells, which are designed not to buckle in a crash. However, the front and back of the cars are designed specifically to buckle in a crash. These are the **crumple zones**. By buckling, the crumple zone reduces the deceleration by increasing the time to slow the passenger. This means that the force acting on the passenger is less.

▲ Figure 5.88

Seat belts

Another vital safety feature in a car is a seat belt – you must wear one by law in most countries. Figure 5.89 shows some statistics that illustrate how wearing seat belts has reduced the rate of serious injuries in car crashes in the UK. In a crash the crumple zone might take a few tenths of a second to buckle and collapse. If you are strapped in with a seat belt, you stop in this time. If you are not strapped in, you keep moving until you hit something in front. Then you stop in a much shorter time and the force stopping you is much greater. Seat belts stretch a little. This allows the passenger some extra time to slow down.

> **TIP**
> A longer time to change momentum means that a smaller force is exerted.

> **KEY TERM**
> **Crumple zones** are zones in the front and back of cars which are designed to buckle in a crash.

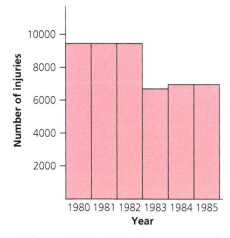

▲ **Figure 5.89** In 1983 a law was passed making it illegal to sit in the front seat of a car without a seat belt. This graph shows the reduction in number of injuries after the introduction of the law.

H

Test yourself

37 Calculate the momentum of:
   a) a car of mass 1200 kg moving at 30 m/s
   b) a 60 kg cyclist pedalling at 8 m/s
   c) a football of mass 0.4 kg moving at 7 m/s
   d) a 500 kg pony galloping at 16 m/s.

38 Using your knowledge of momentum, explain why:
   a) you bend your knees when you land after a jump
   b) cars are designed to have crumple zones
   c) some trainers have gas-filled heels.

   [Hint: in answering this question it can be helpful to use the equation which links force, time and change of momentum.]

39 Explain why passengers should wear seat belts in cars and coaches.

40 Explain why it hurts more if you fall onto a concrete surface than it does if you fall onto a soft grass surface.

◯ Conservation of momentum

Momentum is a very useful quantity when it comes to calculating what happens in collisions. Momentum is always conserved.

When two objects collide, the total momentum they have is the same after the collision as it was before the collision.

This is always true – provided that the colliding objects are in a **closed system**. This means that when the objects collide, no forces outside of the system (external forces) act on the objects.

If an object explodes, the momentum of the object before the explosion is the same as the total momentum of the parts after the explosion. Again, this is always true in a closed system.

Figure 5.90 shows two ice hockey players colliding on the ice. When they meet they push each other. The blue player's momentum decreases and the red player's momentum increases by the same amount.

Before the collision the total momentum was:

$$\text{momentum of blue player} = 100 \times 5 = 500\,\text{kg m/s}$$
$$\text{momentum of red player} = 80 \times 3 = 240\,\text{kg m/s}$$

This makes a total of 740 kg m/s.

> **TIP**
> Momentum is always conserved in a closed system of colliding particles.

> **KEY TERM**
> A **closed system** has no external forces acting on it.

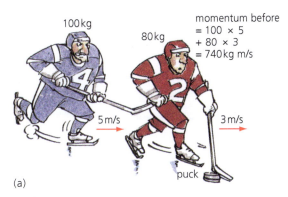

100 kg

80 kg

momentum before
= 100 × 5
+ 80 × 3
= 740 kg m/s

5 m/s

3 m/s

puck

(a)

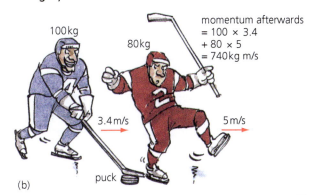

100 kg

80 kg

momentum afterwards
= 100 × 3.4
+ 80 × 5
= 740 kg m/s

3.4 m/s

5 m/s

puck

(b)

▲ **Figure 5.90**

After the collision the total momentum was:

momentum of blue player = 100 × 3.4

= 340 kg m/s

momentum of red player = 80 × 5

= 400 kg m/s

This makes a total of 740 kg m/s which is the same as before.

Momentum as a vector

Figure 5.91 shows a large estate car colliding head-on with a smaller car. They both come to a halt. How is momentum conserved here?

Momentum is a vector quantity. In Figure 5.91a) you can see that the red car has a momentum of +15 000 kg m/s and the blue car has momentum of –15 000 kg m/s. So the total momentum is zero before the collision and there is as much positive momentum as negative momentum. After the collision both cars have stopped moving and the total momentum remains zero.

(a)

total momentum before = 0

+ 10 m/s 1500 kg   − 30 m/s 500 kg

+ 15 000 kg m/s   − 15 000 kg m/s

total momentum after = 0

(b)

1500 kg   500 kg

both stationary

▲ Figure 5.91

Practical

Investigating collisions

Figure 5.92 shows how you can investigate conservation of momentum using laboratory trolleys. One trolley is pushed into a second stationary trolley and the speeds are measured before and after the collision. These speeds can be measured with light gates and data loggers, or with ticker tape.

In Figure 5.92a) a trolley of mass 1 kg, travelling at 1 m/s, collides with an identical trolley, which is at rest. They stick together. The momentum before the collision is:

1 kg × 1 m/s = 1 kg m/s

The trolleys move off after the collision with the same total momentum of 1 kg m/s, but this momentum is shared by the two trolleys, so they move with a speed of 0.5 m/s.

What happens in Figure 5.92b)?
The total momentum before the collision is:

1 kg × 3 m/s = 3 kg m/s

The momentum after the collision = 3 kg m/s (momentum is conserved). The trolleys stick together to make a new mass of 3 kg, which now carries the momentum. So:

3 kg m/s = 3 kg × v

The unknown speed, v, is therefore 1 m/s.

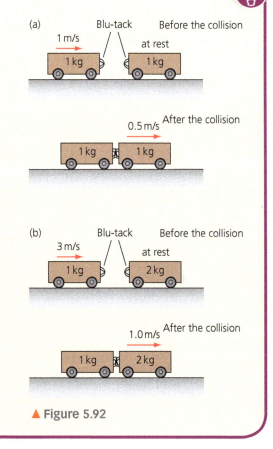

(a) Blu-tack   Before the collision

1 m/s   at rest

1 kg   1 kg

0.5 m/s   After the collision

1 kg   1 kg

(b) Blu-tack   Before the collision

3 m/s   at rest

1 kg   2 kg

1.0 m/s   After the collision

1 kg   2 kg

▲ Figure 5.92

Test yourself

41 In each of the experiments shown in Figure 5.93, the two trolleys collide and stick together. Work out the speeds of the trolleys after their collisions.

42 In Figure 5.91 two cars collided head-on. Each driver had a mass of 70 kg.

   a) Calculate the change of momentum for each driver.

   b) The cars stopped in 0.25 s. Calculate the average force that acted on each driver.

   c) Explain which driver is likely to be more seriously injured.

43 A lorry of mass 18 000 kg collides with a stationary car of mass 2000 kg.

   a) Calculate the momentum of the lorry before the collision.

   b) After the collision the lorry and car move off together. What is their combined momentum after the collision?

   c) What is the speed of the lorry and car after the collision?

   d) Each of the drivers of the car and lorry has a mass of 80 kg. Calculate their changes of momentum during the collision.

   e) The collision lasts for 0.2 s. Calculate the force that acts on each driver during the collision. Explain which driver is more likely to be injured.

   f) Explain why a crumple zone protects the drivers in a collision.

   g) i) Explain why a seat belt helps prevent injury in the event of a crash.

      ii) Explain why seat belts are designed to stretch a little.

44 Use the principle of conservation of momentum to explain the following:

   a) A gun recoils backwards when fired.

   b) You can swim through water.

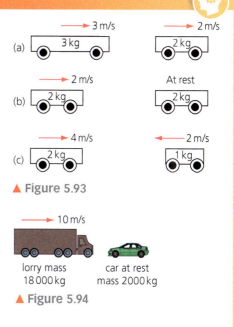

(a) 3 m/s — 3 kg | 2 m/s — 2 kg

(b) 2 m/s — 2 kg | At rest — 2 kg

(c) 4 m/s — 2 kg | 2 m/s — 1 kg

▲ **Figure 5.93**

10 m/s

lorry mass 18 000 kg   car at rest mass 2000 kg

▲ **Figure 5.94**

Show you can...

Show you understand the principle of the conservation of momentum by completing this task.

Two trolleys, which are travelling in opposite directions, collide and stick together. Explain how you can predict their motion after the collision, if you know their masses and velocities before the collision.

Calculations using two equations

Sometimes to calculate the quantity you have been asked to find you need to use two equations.

Example

A racehorse accelerates out of the starting gate at 7.2 m/s^2. The racehorse maintains this acceleration for a distance of 22.5 m.

Calculate the momentum of the racehorse at this point in the race.

The mass of the racehorse is 800 kg.

Answer

The data given in the question:

acceleration (a) = 7.2 m/s^2

distance (s) = 22.5 m

mass (m) = 800 kg

initial velocity (u) = 0 m/s (the horse was not moving at the start of the race)

To calculate v:

$$v^2 - u^2 = 2as$$

$$v^2 - 0^2 = 2 \times 7.2 \times 22.5$$
$$v^2 = 324$$
$$v = \sqrt{324} = 18 \text{ m/s}$$

Now use the equation:

momentum = mass × velocity

$$\text{momentum} = 800 \times 18 = 14\,400 \text{ kg m/s}$$

Chapter review questions

1   Figure 5.95 shows a distance–time graph for a car on a motorway.

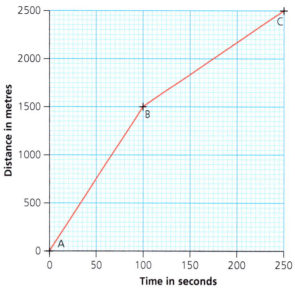

▲ Figure 5.95

a) Where is the car travelling faster, A to B or B to C?

b) i) How long did it take to travel from A to B?

   ii) How far did the car travel from A to B?

   iii) How far did the car travel from B to C?

c) Calculate the speed of the car between A and B.

2   A train travels at a constant speed of 45 m/s. Calculate:

a) how far it travels in 30 s

b) the time it takes to travel 9 000 m.

3   A car has a mass of 1 500 kg. It takes 6 s to increase its velocity from 5 m/s to 23 m/s.

a) i) Calculate the change of velocity.

   ii) Calculate the car's acceleration.

b) When travelling at 23 m/s the driver takes his foot off the accelerator. The car takes 20 s to slow down to 15 m/s.

   i) Calculate the deceleration of the car.

   ii) Calculate the air resistance acting on the car to slow it down.

4   Explain why a skydiver wearing loose clothing and spreading out his arms and legs has a lower terminal velocity than a skydiver curled up into a ball.

5 Figure 5.96 shows a velocity–time graph for an aeroplane just before it takes off.

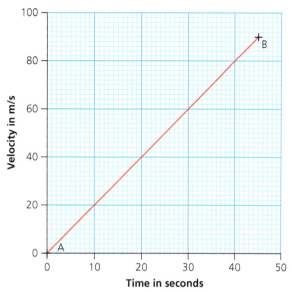

▲ Figure 5.96

a) Calculate the acceleration of the aeroplane.

b) Calculate the distance covered by the plane before it takes off at time B.

6 Calculate the work done in each of the following cases.

a) A car is accelerated by a force of 400 N over a distance of 80 m.

b) A girl with a weight of 470 N climbs a staircase with a height of 3.6 m.

c) You hold a weight of 200 N stationary over your head for 5 seconds.

d) A boat moves a distance of 3 000 m when a drag force of 60 000 N acts on it.

7 A skydiver with a weight of 700 N falls at a constant speed.

a) What is the drag force that acts on her?

b) When she opens the parachute, a drag force of 1 500 N acts on her. Explain what happens to her speed.

8 In a car crash, it takes about 0.04 s for an air bag to inflate. A fraction of a second later, the driver's head hits the air bag. The bag, which has tiny holes in it, starts to deflate as the impact of the driver's head pushes some gas out.

a) What happens to the movement of the driver's head when it hits the air bag?

b) Why is it important for the air bag to deflate?

c) Explain how the air bag reduces the risk of serious injury.

9 A tank moving at 6.5 m/s has a momentum of 195 000 kg m/s. Calculate the mass of the tank.

10 Emma test drives her new car. She accelerates rapidly reaching a speed of 20 m/s after 8 seconds. The car and Emma have a combined mass of 800 kg.

a) Calculate the acceleration of the car.

b) Calculate the accelerating force on the car.

c) On another day Emma gets three friends into the car to show off the car's acceleration.

   i) Explain why the car accelerates less rapidly.

   ii) Emma's friends have a combined mass of 450 kg. Calculate the acceleration of the car now. Assume the car experiences the same accelerating force. ▶

extra case added to increase the weight

▲ Figure 5.97

11 A student investigated how the weight of a parachute affects how fast the parachute falls. The student used a cupcake case as the parachute. He kept the area of the case constant, but varied the weight by adding extra cases inside each other.

The table shows the time taken by each set of cases to fall a height of 4 m. The student measured the time of fall for each set of cases three times.

Number of cake cases	Time of fall in s
1	2.7, 2.6, 2.6
1.5	2.2, 2.3, 2.2
2	2.0, 2.0, 1.9
3	1.5, 1.6, 1.7
4	1.4, 1.4, 1.4
6	1.3, 1.3, 1.2
8	1.1, 1.1, 1.2
10	1.1, 1.1, 1.0

a) Why did the student measure the time taken to fall three times for each set of cake cases?

b) Copy the table and add two further columns to show:

 i) the average time of fall

 ii) the average speed of fall.

c) Plot a graph of the number of cases on the *y*-axis against the average speed of fall on the *x*-axis. Draw a line of best fit through the points and comment on any anomalous results.

d) Use the graph to predict the speed of fall for a weight of seven cases.

e) What can you conclude from the data displayed in the graph?

f) Assuming that the cases reach their terminal velocity quickly after release, comment on how the resistive force on the falling cupcake cases depends on their speed.

12 This is a question for some research; you can look on the web, or you might find some useful information from your parents or grandparents.

Driving is much safer now than it was 80 years ago. In the 1930s there were about 2.5 million cars on our roads and over 7000 people died each year in road accidents. Now there are over 32 million cars on the road and on average 2000 people die each year. While this is still a sad statistic, we are much safer now.

Write a paragraph to explain what measures have been put in place over the last 80 years to improve road safety.

Practice questions

1 A triathlon race has three parts: swimming, cycling and running.

Figure 5.98 shows the force responsible for the forward movement of the athlete in the swimming race.

▲ **Figure 5.98**

a) Copy Figure 5.98 and show three other forces acting on the athlete. [3 marks]

b) The table shows the distance of each part of a triathlon race and the time an athlete takes for each part.

Part of race	Distance in m	Time in s
Swimming	1500	1200
Cycling	40 000	3600
Running	10 000	2000

i) Calculate the athlete's average speed during the swim. [1 mark]

ii) Calculate the athlete's average speed for the whole race. [2 marks]

iii) The graph shows how the distance varied with time for the running part of the race.

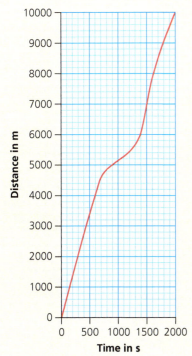

▲ **Figure 5.99**

c) Describe how the athlete's speed changed during this part of the race. [3 marks]

2 The table gives information about a journey made by a cyclist.

Time in hours	Distance in km
0	0
1	15
2	30
3	45
4	60
5	75
6	90

a) Plot a graph using the data in the table. [3 marks]

b) i) Use your graph to find the distance in kilometres that the cyclist travelled in 4.5 hours. [1 mark]

ii) Use the graph to find the time in hours taken by the cyclist to travel 35 kilometres. [1 mark]

c) Write down the equation which links average speed, distance moved and time taken. [1 mark]

3 A train travels between two stations. The velocity–time graph in Figure 5.100 shows the train's motion.

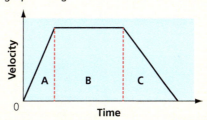

▲ **Figure 5.100**

a) How do you know that the train is decelerating in part C? [1 mark]

b) State the features of the graph that represent the distance travelled between the two stations. [1 mark]

c) A second train travels between the two stations at a constant velocity and does not stop. It takes the same time as the first train. Draw a copy of the graph and then draw a line showing the motion of the second train. [2 marks]

4 a) A lorry is travelling in a straight line and it is accelerating. The total forward force on the lorry is *F* and the total backward force is *B*.

i) Which is larger, force *F* or force *B*? Give a reason for your answer. [1 mark]

ii) Write down the equation that links acceleration, mass and resultant force. [1 mark]

iii) A resultant force of 15 000 N acts on a lorry. The mass of the lorry is 12 500 kg. Calculate the lorry's acceleration and give the unit. [3 marks]

b) The thinking distance is the distance a vehicle travels in the driver's reaction time. The braking distance is the distance a vehicle travels when the brakes are applied.

i) State one factor that increases the thinking distance. [1 mark]

ii) State one factor that increases the braking distance. [1 mark]

c) A council decides to impose a 20 mph speed limit in a town centre. A councillor says than drivers can react more quickly at 20 mph than they can at 30 mph.

What is wrong with this statement? [2 marks]

5 Figure 5.101 shows the minimum stopping distances, in metres, for a car travelling at different speeds on a dry road.

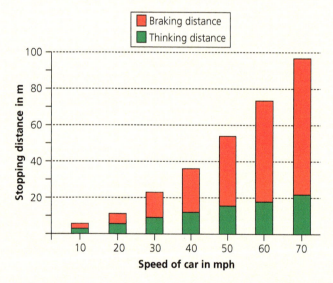

▲ Figure 5.101

a) Write an equation which links stopping distance, thinking distance and braking distance. [1 mark]

b) Describe the patterns shown in the graph. [2 marks]

c) Use the graph to estimate the stopping distance for a car travelling at 35 miles per hour. [1 mark]

d) To find the minimum stopping distance, several different cars were tested. Suggest how the data from the different cars should be used to give the values in the graph. [1 mark]

e) The tests were carried out on a dry road. If the road was icy, describe and explain what change if any there would be to:

i) the thinking distance [2 marks]

ii) the braking distance. [2 marks]

6 Figure 5.102 shows how the velocity of an aircraft changes as it accelerates along a runway. The aircraft takes off after 60 seconds.

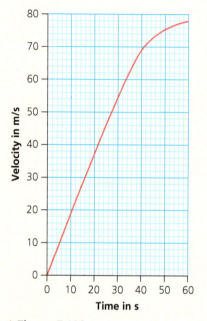

▲ Figure 5.102

a) Use the graph to find the average acceleration of the aircraft over the 60 seconds. [3 marks]

b) Explain why the acceleration is not constant, even though the engines produce a constant force. [3 marks]

c) Use the graph to estimate the length of the runway. [3 marks]

7 A cyclist is travelling at 3 m/s and in 8 seconds increases his speed to 7 m/s.

a) Calculate his acceleration. [3 marks]

b) The mass of the cyclist and his bicycle is 80 kg. The force forwards on the bicycle is 60 N. Calculate the size of air resistance, R, acting on him. Assume he has the acceleration calculated in part (a). [3 marks]

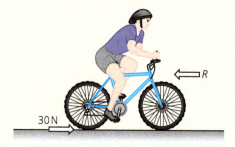

▲ Figure 5.103

8 Scientists test the safety features of a car by crashing it into a very large block of concrete. A dummy is placed in the driver's seat and the scientists video the crash.

▲ Figure 5.104

a) In one test, the dummy and the car travel at 6 m/s. The mass of the dummy is 68 kg. Calculate the momentum of the dummy. [2 marks]

b) In another test, the momentum of the dummy changes by 840 kg m/s in a time of 0.14 s. Calculate the average horizontal force acting on the dummy during this time. [2 marks]

c) These tests help to make our roads safer.

i) State **two** factors that affect the stopping distance of a car on the road. [2 marks]

ii) Use ideas about momentum to explain how crumple zones help reduce injuries during a crash. [2 marks]

9 A group of students uses a special track. The track is about 2 metres long and is horizontal. Two gliders, P and Q, can move along the track.

The surface of the track and the inside surface of the gliders are almost frictionless. Figure 5.105 shows that the gliders can move through two light gates, A and B.

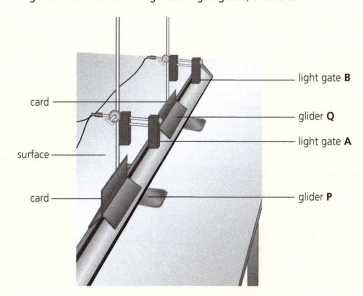

▲ Figure 5.105

The mass of glider P is 2.4 kg. This glider is moving towards Q at a constant velocity of 0.6 m/s. Glider Q is stationary.

Figure 5.106 shows a side view. Each glider has a card and a magnet attached. Light gate A records the time for which the card is in front of the light gate.

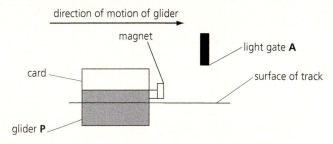

▲ Figure 5.106

a) i) Apart from the time recorded by the light gate A, state the other measurement that would be needed to calculate the velocity of glider P. [1 mark]

ii) Why does the surface of the track need to be frictionless and horizontal? [1 mark]

b) Momentum is a vector quantity.

i) State what is meant by a vector quantity. [1 mark]

ii) Calculate the momentum of glider P. [2 marks]

iii) State the momentum of glider Q. [1 mark]

c) Glider P collides with glider Q and they move off together at a speed of 0.4 m/s.

i) State the combined momentum of P and Q after the collision. [1 mark]

ii) Calculate the mass of glider Q. [2 marks]

iii) Calculate the change in momentum of glider P during the collision. [2 marks]

iv) The time taken for the collision was 0.05 s. Calculate the force that acted on glider P during the collision. [3 marks]

v) State the size of the force acting on glider Q during the collision. [1 mark]

177

Working scientifically

Independent, dependent and control variables

To investigate the idea of a crumple zone, Tracy used the apparatus shown in Figure 5.107 (She put some crumpled waste paper down to catch the weight too). Tracy expected that in a collision the test material would slow down the trolley in the same way as the crumple zone slows down a car.

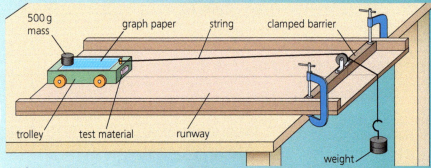

▲ Figure 5.107

A constant force was applied to the trolley by a falling weight. When the trolley hit the barrier, it decelerated and stopped. The 500g mass on top of the trolley slid forwards until it also stopped. Tracy marked the graph paper to show how far the 500g mass moved. Tracy tested five different types of material for the crumple zone, each of the same area and thickness.

1 Why does the 500g mass slide forwards when the trolley hits the barrier?

2 In this investigation, what is:

  a) the independent variable

  b) the dependent variable?

Remember that the independent variable is the variable that affects the dependent variable.

By applying the same force to the trolley and starting it from the same position on the runway, the trolley always hit the barrier at the same speed. The speed that the trolley hits the barrier is a control variable.

3 What else was a control variable in this investigation?

> **KEY TERMS**
>
> An **independent variable** is the variable that you change. An investigation should only have one independent variable.
>
> A **dependent variable** is the variable that changes because of a change made to the independent variable.
>
> A **control variable** is what you keep the same.

Figure 5.108 shows the data Tracy obtained using two different accelerating forces. The lines can be used to measure how far the 500 g mass moves forwards mass for different types of test material.

(a)

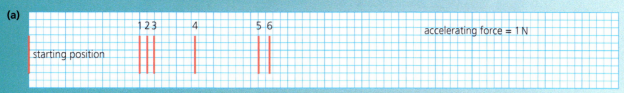

(b)

Key

1 = no material; 2 = polystyrene; 3 = foam rubber; 4 = carpet; 5 = fibre wool; 6 = rubber carpet underlay

▲ Figure 5.108

4 What **range** of distances did the 2 N accelerating force give?

5 Draw a bar chart for each set of data.

6 Why is it more appropriate to draw a bar chart than a line graph?

7 Are the patterns shown by the data consistent? Give a reason for your answer.

8 Tracy only obtained one set of data for each accelerating force.

Why might it have been better had Tracy obtained two sets of data for each force?

In her analysis of the investigation, Tracy wrote:

'The model crumple zones did reduce the force on the trolley by increasing the time it took the trolley to stop. Of the five materials that I tested the best was the rubber carpet underlay.'

9 Do both sets of data support Tracy's analysis? Explain the reasons for your answer.

KEY TERM

Range refers to the maximum and minimum values used or recorded.

6 Waves

Electromagnetic waves, which travel at the speed of light, allow us to communicate with friends all around the world. These waves transfer energy for receivers to detect them, and information for us to understand the message.

Specification coverage

This chapter covers specification points: 4.6.1 Waves in air, fluids and solids, 4.6.2 Electromagnetic waves and 4.6.3 Black body radiation.

Previously you could have learned:

› **Sound waves are mechanical waves which travel through a medium such as air or water.**

› **Light waves are examples of electromagnetic waves which travel very quickly. These waves are able to travel through a vacuum as well as media such as air or glass.**

When you drop a stone into a pond you see water ripples spreading outwards from the place where the stone landed (with a splash). As the ripples spread, the water surface moves up and down. These ripples are examples of **waves**.

The water ripples transfer energy and information. The energy moves outwards from the centre but the water itself does not move outwards. The shape of the waves provides us with the information about where the stone landed (if we did not see it land).

Test yourself on prior knowledge

1 Give three examples of waves which have not already been mentioned in the section above.
2 For each of your examples in Question 1, explain how the wave transfers energy and information.
3 How do thunderstorms help us to realise that light and sound transfer at different speeds?

▲ Figure 6.1

Waves in air, fluids and solids

○ Transverse and longitudinal waves

A good way for us to visualise waves is to use a stretched 'slinky' spring. When two students stretch a slinky across the floor in a laboratory, they can transmit waves that travel slowly enough for us to see.

KEY TERM

A **transverse wave** is one in which the vibration causing the wave is at right angles to the direction of energy transfer.

Transverse waves

Figure 6.2 shows a **transverse wave**. One student holds the end of the slinky stationary. The other end of the slinky is moved from side to side. A series of pulses moves down the slinky, sending energy from one end to the other. The student holding the slinky still feels the energy as it arrives. None of the material in the slinky has moved permanently.

Water waves and light waves are two examples of transverse wave. The water shown in Figure 6.1 moves up and down. Energy is carried outwards by the wave, but water does not pile up at the edge of the pond.

direction of energy transfer

movements of hand from side to side

the tape moves from side to side

this end is held still

▲ **Figure 6.2** The transverse waves transfer energy along the slinky from one end to the other. The coloured tape shows that the pulses cause the slinky to vibrate from side to side, just the same as the student's hand.

KEY TERM ⭐

A **longitudinal wave** is one in which the vibration causing the wave is parallel to the direction of energy transfer.

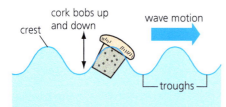

▲ **Figure 6.3** Water waves cause a cork to bob up and down, as the peaks and troughs of the wave pass.

▶ **Figure 6.5** Earthquakes produce shock waves, which may be either longitudinal or transverse waves. Buildings are damaged by the transfer of energy, not by the movement of material.

Show you can…

A friend of yours has been away and missed the first lesson on waves. Your friend thinks that water waves are carried by the sea moving forwards.

Explain how we know that waves carry energy without the sea (or any other medium) moving along with the waves.

▲ **Figure 6.6**

Longitudinal waves

Figure 6.4 shows a **longitudinal wave**. Energy is transmitted along the slinky by pulling and pushing the slinky backwards and forwards. This makes the slinky vibrate backwards and forwards. The vibration of the slinky is parallel to the direction of energy transfer. The coils are pushed together in some places (areas of **compression**). In other places the coils are pulled apart (areas of **rarefaction**). As energy is transferred, none of the material of the slinky moves permanently.

Sound is an example of a longitudinal wave. When a guitar string is plucked, energy is transferred through the air as the string vibrates backwards and forwards. However, the air itself does not move away from the string with the wave – there is not a vacuum left near the guitar.

▲ **Figure 6.4** The coloured tape on the slinky shows that the pulses on the slinky move the coils backwards and forwards in the same direction as the student's hand moves.

Test yourself

1 Use diagrams to illustrate the nature of
   a) transverse waves
   b) longitudinal waves.
2 What do the terms 'area of compression' and 'area of rarefaction' mean?
3 Lucy is playing in her paddling pool. As she pushes a ball up and down in the water it makes a small wave (Figure 6.6).
   a) Which way does the water move as the waves move across the pool?
   b) Describe the motion of the other balls in the pool.
4 Describe how you would use a slinky to show that waves transfer both energy and information.

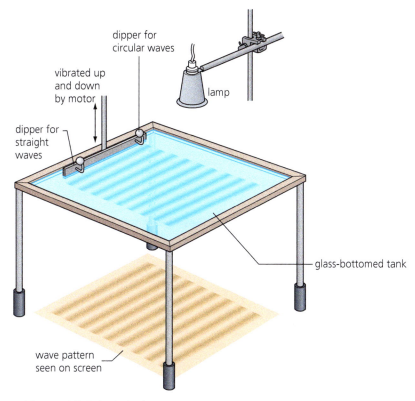

▲ **Figure 6.7** A ripple tank.

◯ **Properties of waves**

We can learn more about the properties of waves by looking at water waves in a ripple tank as shown in Figure 6.7.

We can produce waves by lowering a dipper into the tank. A wooden bar is used to produce straight waves (plane waves) and a spherical dipper produces circular waves. By shining a light from above, we see the pattern of waves produced by the peaks and troughs of the waves (Figure 6.7).

Describing waves

Waves are described by the terms, **amplitude** (*A*), **wavelength** (*λ*), **frequency** (*f*) and **period** (T). *λ*, the symbol for wavelength, is a Greek letter pronounced *lambda*.

Figure 6.8 shows a transverse wave. This diagram helps us to understand the meaning of these terms.

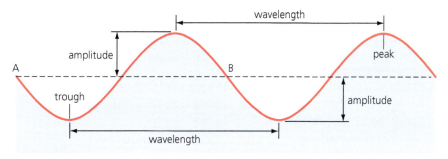

▲ **Figure 6.8**

- In Figure 6.8 the amplitude is the distance from a peak to the middle or from a trough to the middle.
- In Figure 6.8 a wavelength, *λ*, is the distance between two peaks or two troughs.
- Frequency is measured in hertz (Hz). A frequency of 1 Hz means a source is producing one wave per second. If a student, using a slinky, moves his hand from side to side and back twice each second, he produces two complete waves each second. The frequency of the waves is 2 Hz.
- The period, *T*, of a wave is the time taken to produce one wave.
- The frequency and time period of a wave are linked by this equation:

$$T = \frac{1}{f}$$

$$\text{period} = \frac{1}{\text{frequency}}$$

where period is in seconds, s
      frequency is in hertz, Hz.

KEY TERMS ⭐

Amplitude is the height of the wave measured from the middle (the undisturbed position of the water).

Wavelength is the distance from a point on one wave to the equivalent point on the next wave.

Frequency is the number of waves produced each second. It is also the number of waves passing a point each second.

Period is the time taken to produce one wave.

Example

Calculate the period of a wave with a frequency of 10 Hz.

$$T = \frac{1}{10}$$
$$= 0.1\,s$$

You can remember this as follows: if a source produces 10 waves each second, the source takes 0.1 s to produce one wave.

Example

Sound travels at a speed of 330 m/s in air. Calculate the wavelength of a sound wave with a frequency of 660 Hz.

$$v = f\lambda$$
$$330 = 660 \times \lambda$$
$$\lambda = \frac{330}{660}$$
$$= 0.5\,m$$

Wave speed and the wave equation

The **wave speed** is the speed at which energy is transferred (or the wave moves) through the medium.

All waves obey this equation:

$$v = f\lambda$$

wave speed = frequency × wavelength

where wave speed is in metres per second, m/s

frequency is in hertz, Hz

wavelength is in metres, m.

Test yourself

5 Figure 6.9 shows wave pulses travelling at the same speed along two ropes. How are the wave pulses travelling along rope A different from the wave pulses travelling along rope B?

6 Figure 6.10 shows transverse waves on a rope of length 8 m.
   a) What is the name given to each of these horizontal distances?
      i) ae   ii) bf   iii) dg
   b) What is the name given to each of these vertical distances?
      i) ed   ii) ef
   c) Calculate:
      i) the wavelength of the wave ii) the amplitude of the wave
      iii) the horizontal distance ag.

7 The periods of two waves are:
   a) 0.25 s   b) 0.01 s.
   Calculate the frequency of each wave.

8 A stone is thrown into a pond and waves spread outwards. The waves travel with a speed of 0.4 m/s and their wavelength is 8 cm. Calculate the frequency of the waves.

9 Make a sketch to show the wavelength of a longitudinal wave.

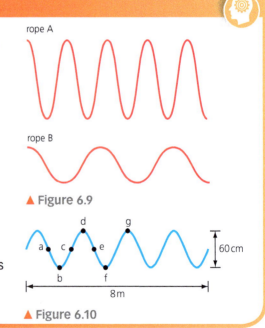

▲ Figure 6.9

▲ Figure 6.10

Show you can...

Show you understand the key words which describe waves by making a sketch which shows the wavelength and amplitude of a transverse wave.

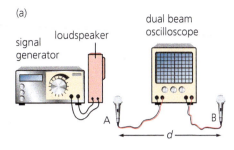

(a)

signal generator | loudspeaker | dual beam oscilloscope

A   B

d

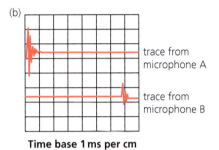

(b)

trace from microphone A

trace from microphone B

Time base 1 ms per cm

▲ Figure 6.11

○ Measuring wave speeds

Three different methods of measuring wave speeds are described below.

Measuring the speed of sound in a laboratory

It is possible to measure wave speed using the equation:

$$\text{speed} = \frac{\text{distance travelled}}{\text{time taken}}$$

The problem with measuring the speed of sound in air is that sound travels quickly. So we must find a way to measure short times accurately in the laboratory. Figure 6.11 shows how we can do it.

- A loudspeaker is connected to a signal generator which produces short pulses of sound.
- Two microphones are placed near the loudspeaker but separated by a short distance, *d*. Each microphone is connected to an input of a dual beam oscilloscope.
- The oscilloscope can measure the time difference, *t*, between the sound reaching microphone A and microphone B.

Example

A student sets the microphones a distance of 220 cm apart. The time base on the oscilloscope is set to 1 ms per cm: this means that a distance of 1 cm on the horizontal scale corresponds to a time of 1 ms (0.001 s).

Using Figure 6.11b) you can measure that the sound reaches microphone B 6.6 ms after the sound reaches microphone A.

$$\begin{aligned}\text{So speed of sound} &= \frac{d}{t}\\ &= \frac{2.2}{0.0066}\\ &= 330\,\text{m/s}\end{aligned}$$

Remember: you must convert distance into metres and time into seconds.

Measuring the speed of sound outside

- When outside, the echoes from a tall building can be used to measure the speed of sound in air.
- Stand 40 m in front of a tall building and bang two blocks of wood together.
- Each time you hear an echo, bang the blocks together again.
- Have another student use a stopwatch to time how long it took to hear 10 echoes.

The results obtained by two students are given in the table.

▲ Figure 6.12 Using echoes to measure the speed of sound in air.

Trial number	Time in seconds
1	2.4
2	2.8
3	2.3

1 Calculate the mean value of the time for 10 echoes.
2 How long does it take the sound to travel 80 m?
3 Calculate the speed of sound given by these results.

Investigating waves in a ripple tank and waves in a stretched string
Waves in a ripple tank

Method
1 Set up the ripple tank as shown in Figure 6.7.
2 Pour enough water to fill the tank to a depth of about 5 or 6 mm.
3 Adjust the wooden bar up or down so that it just touches the surface of the water.
4 Switch on the lamp and the electric motor.
5 Adjust the speed of the motor so that low frequency waves that can be counted are produced.
6 Move the lamp up or down so that a clear pattern can be seen on the floor.
7 If the pattern is difficult to see a sheet of white paper or card on the floor under the tank and the laboratory lights off may help.
8 Use a metre ruler to measure across as many waves in the pattern as possible. Divide that length by the number of waves. This gives the wavelength of the waves.
9 Count the number of waves passing a point in the pattern over a given time (say 10 seconds).

Divide the number of waves counted by 10. This gives the frequency of the waves.

If you have connected the motor to a variable frequency power supply you can take the frequency directly from the power supply.

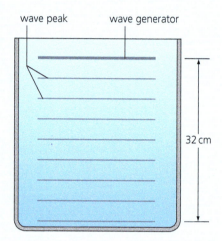

▲ Figure 6.13 Eight wavelengths cover a distance of 32 cm so one wavelength is equal to 32 ÷ 8 = 4 cm.

Analysing the results
Use the equation:

$$\text{wave speed} = \text{frequency} \times \text{wavelength}$$

and your values for wavelength and frequency to calculate the wave speed.

Waves in a stretched string

Figure 6.14 shows how the waves in a stretched string can be investigated. You can use either a string or an elastic cord for this investigation.

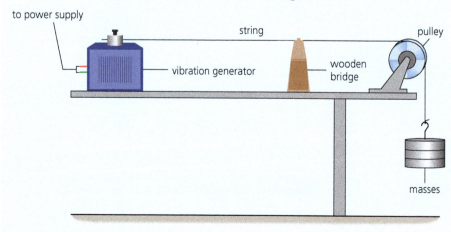

▲ Figure 6.14

Method

1 Switch on the vibration generator. The string (or elastic cord) will start to vibrate but not in any pattern.

2 Changing the mass attached to the string changes the tension in the string. Moving the wooden bridge changes the length of the string that vibrates.

3 Change the mass or move the wooden bridge until you see a clear wave pattern. The pattern will look like a series of loops. The length of each loop is half a wavelength.

4 Use a metre ruler to measure across as many loops as possible. Divide this length by the number of loops then multiply by two. Your final answer is equal to the wavelength of the wave.

5 The frequency of the wave is the same as the frequency of the power supply.

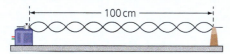

▲ **Figure 6.15** There are 10 loops in a distance of 100 cm. So the length of each loop is 10 cm making the wavelength of the wave equal to 20 cm.

Analysing the results

Use the equation:

$$\text{wave speed} = \text{frequency} \times \text{wavelength}$$

and your values for wavelength and frequency to calculate the wave speed.

Take it further

Use the same apparatus to investigate how the velocity of a wave on a stretched string depends on the tension in the string. Identify suitable apparatus for making each of the measurements that will need to be taken.

Test yourself

10 Describe how you would measure the wave speed on a stretched slinky spring which is about 5 m long.

11 A student is using the apparatus in Figure 6.11. She moves the microphones closer together so that the horizontal distance between the traces, on the oscilloscope, from the two microphones is 4.2 cm. Calculate how far apart the microphones are. (The speed of sound in air is 330 m/s.)

Waves travelling from one medium to another

Figure 6.16 shows a ripple tank viewed from the top. The black lines represent peaks of waves. In the bottom right-hand part of the tank, there is a region of shallow water. Water waves travel more slowly in shallow water. You can see that the wavelength of the waves also decreases.

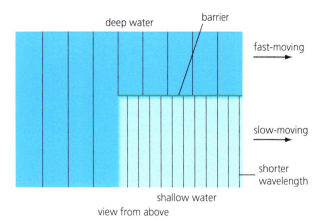

▲ Figure 6.16

When waves travel from one medium to another the frequency stays the same.

The speed and wavelength of the waves are related by the equation:

$$v = f\lambda$$

So when waves travel from medium A to medium B: the wavelength in medium A is longer than in medium B, when the speed in medium A is greater than the speed in medium B.

This rule holds true for all types of wave including sound waves and light waves.

Test yourself

12 Figure 6.17 shows sound waves travelling in air at a speed of 330 m/s. The sound passed through a thin rubber surface into a balloon of gas.

a) What happens to the frequency of the waves as they pass into the gas?

b) Draw diagrams to show what happens to the wavelength of the sound when the balloon is filled with:

i) carbon dioxide in which sound travels at 270 m/s

ii) neon in which sound travels at 460 m/s.

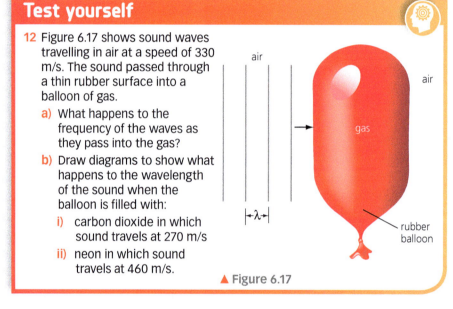

▲ Figure 6.17

▲ Figure 6.18

○ Reflection

Mirrors work by reflecting light. A mirror with a flat surface is called a **plane mirror**. Figure 6.19 shows how you can investigate the reflection of light using a ray box and a plane mirror. The ray box produces a narrow beam of light. This is called a **ray**.

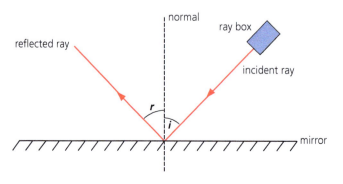

▲ Figure 6.19

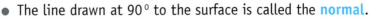

KEY TERMS

The **normal** is a line drawn at 90° to a surface where waves are incident.

The **angle of incidence** is the angle between the incident ray and the normal.

The **angle of reflection** is the angle between the reflected ray and the normal.

- The line drawn at 90° to the surface is called the **normal**.
- The angle between the incident ray and the normal is called the **angle of incidence**.
- The angle between the reflected ray and the normal is called the **angle of reflection**.

Whenever light is reflected, the ray obeys this law:

angle of incidence, i = angle of reflection, r

This law of reflection is true for all types of wave including sound waves and water waves. Figure 6.20 shows water waves being reflected off a solid barrier in a ripple tank.

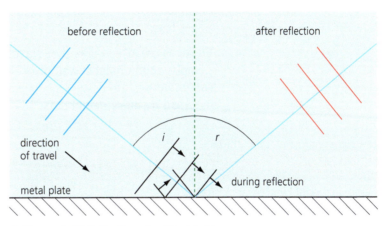

▲ **Figure 6.20** Reflection of water waves.

Images in a mirror

Figure 6.21 is a ray diagram to show how an image is formed by a plane mirror. Two rays from the object (the red and the green ray) are reflected by the mirror into an eye. These rays are diverging and appear to have come from a point behind the mirror. The brain interprets the rays as coming from an image behind the mirror. This is called a **virtual image**. It is not really there. The virtual image is the right way up. When you look at yourself in a mirror and raise your right hand, your image raises its left hand. The ray diagram shows this: when you look towards the object the green ray is on the left; when you look at the image the green ray is on the right.

A plane mirror always produces an upright virtual image.

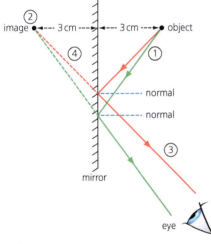

▲ **Figure 6.21**

TIP

In your exam, you may be given a diagram with the object and be asked to complete the diagram to show how an image is formed. You need to practise doing this. Follow the procedure as shown in Figure 6.21:

1 Draw two rays from the object – the green and red lines.
2 Mark in the position of the image – you know it must be behind the mirror at the same distance as the object.
3 Draw the reflected rays, so that they point towards the image.
4 Add in the dotted 'virtual' lines to the image.

KEY TERMS

A **virtual image** is an image formed by light rays which appear to diverge from a point.

A medium which **transmits** light allows the light to pass through it.

A medium which **absorbs** light does not allow light to pass through it.

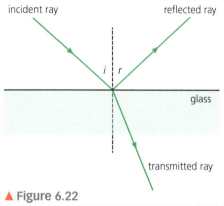

incident ray   reflected ray

i | r

glass

transmitted ray

▲ **Figure 6.22**

Transmission and absorption of waves

A mirror is a very good reflector of light. Nearly all the light that is incident on the mirror is reflected. Some other surfaces either **transmit** or **absorb** light.

Figure 6.22 shows a light ray incident on the surface of a block of glass. Some of the light is reflected off the surface, but some of the light travels into the glass. This is the transmitted ray. You can see that the transmitted ray changes direction. This usually happens when waves travel from one medium to another.

Some surfaces absorb light. Most houses have ceilings painted white and walls painted in light colours, so that rooms are bright and well lit. If you put black or dark coloured paint on the wall, the room will be darker. This is because the dark coloured paints absorb light and reflect less.

All waves are affected as they travel from one medium to another. For example, sound reflects well off a cliff and you hear echoes in mountains. Modern houses are designed to absorb sound, so that we are not disturbed by noisy neighbours.

Test yourself

13 Describe what is meant by a 'plane mirror'.

14 a) John stands 5 m in front of a plane mirror. Where does his image appear to be?

   b) John walks towards the mirror at 2 m/s. At what speed does his image appear to approach him?

15 A periscope is useful for looking over a crowd. Figure 6.23 shows two plane mirrors used as a periscope.

   a) Copy and complete the diagram to show the path taken by the two light rays from the object to the eye.

   b) Describe the image obtained using the periscope.

16 Copy the diagram shown in Figure 6.24. Complete the path of the ray after reflection. Draw the diagram accurately using a protractor.

17 Figure 6.25 shows a photograph of an acoustic panel. Name two features of the panel which reduce sound transmission. Give reasons for your answer.

▶ **Figure 6.25** The pyramid shapes on this acoustic foam ensure that little sound is transmitted out of a room.

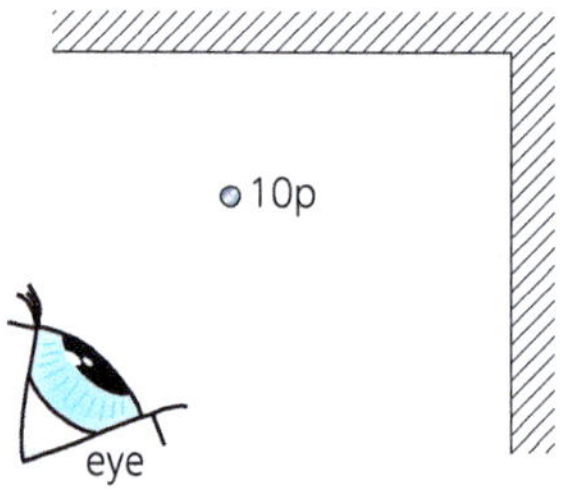

18 a) How many 10p coins will the eye see reflected in the two mirrors (Figure 6.26)?

   b) Draw a diagrams to show the position of the images.

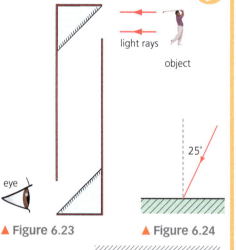

light rays

object

25°

eye

▲ **Figure 6.23**   ▲ **Figure 6.24**

10p

eye

▲ **Figure 6.26**

Show you can...

Show you understand how a mirror forms an image by completing this task.

A small object is placed 10 cm in front of a plane mirror. Draw an accurate diagram to show how an image is formed behind the mirror.

Sound waves

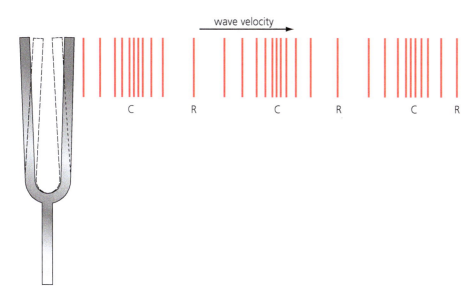

When an object vibrates it can produce a sound. For example, a tuning fork is used to produce a musical note. The prongs of the fork vibrate and produce compressions and rarefactions in the air. Sound is a longitudinal wave.

▲ **Figure 6.27** The vibration of the tuning fork is creating compressions (C) and rarefactions (R).

Sound waves can also travel through solids and fluids by setting up vibrations in those media. You can make a glass 'ring' by rubbing a wet finger round its rim. Vibrations in the glass send sound waves into the air.

Our ears detect sound by converting vibrations in the air (sound waves) into mechanical vibrations in a solid (Figure 6.28).

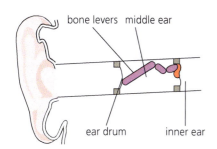

▲ **Figure 6.28** The human ear.

The compressions and rarefactions in the air set the ear drum in motion. As the ear drum vibrates backwards and forwards, small bones are made to vibrate. The vibrating bones are connected to nerves which give us the sensation of sound.

The conversion of sound waves to vibrations in the bones only works over a limited range of frequencies. The human ear can detect sound over a range of 20 Hz–20 000 Hz (20 kHz). Other animals can hear outside this range. Elephants can hear below 20 Hz; dogs and cats up to 40 000 Hz; dolphins and bats can hear up to about 160 000 Hz.

We also detect sounds with microphones. The sound causes a diaphragm vibrate, which produces an electrical signal (see page 235 in Chapter 7).

○ Using waves for detection and exploration

Ultrasound

Ultrasound waves are sound waves which have a frequency higher than the limit of human hearing (20 kHz). Very high frequency ultrasound is used by doctors to explore inside our bodies. The ultrasound is directed towards an organ in a narrow beam (like a light ray). At each surface some of the ultrasound is reflected and some transmitted. The reflections are detected, at different times, and a computer builds up an image of the organ, without the doctor having to look directly. Ultrasound is used for foetal scanning.

One great advantage of ultrasound is that it is safe and does no harm to the patient.

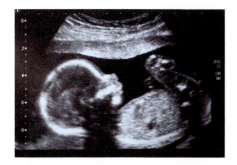

▲ **Figure 6.29** Ultrasound scans are widely used to check on the health of an unborn baby. This is called foetal scanning.

KEY TERM

Ultrasound is a sound wave with a frequency above the range of human hearing (above 20 kHz).

Ships use beams of ultrasound (sonar) for a variety of purposes. Fishing boats look for fish, the depth of shipping lanes can be checked, and naval ships can search for submarines.

Figure 6.30 shows the idea. A beam of ultrasound is transmitted by the ship. The waves are reflected back off the sea bed. The longer the delay between the transmitted and the reflected signals, the greater the depth of the sea.

(b)

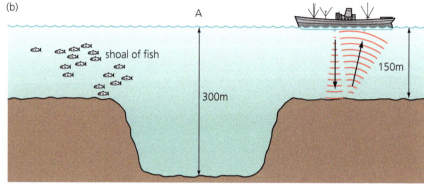

(a)

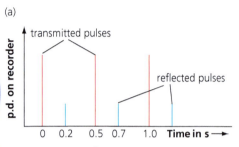

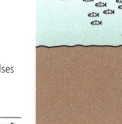

▲ **Figure 6.30** Using sonar.

Seismic waves

KEY TERMS

P-waves are longitudinal seismic waves.

S-waves are transverse seismic waves.

Seismic waves are produced by earthquakes. When we hear about an earthquake, we usually think about the destruction they cause, but earthquakes also give us information about the Earth's structure.

- **P-waves** are longitudinal seismic waves. P-waves travel at different speeds through solid and liquid rock.
- **S-waves** are transverse seismic waves. S-waves can only travel through solid rock, they cannot travel through liquid.

The Earth has three major layers.

- The outer layer is the mantle; this is solid. P-waves and S-waves can travel through the mantle.
- The outer core is liquid. Only P-waves can travel through the outer core.
- The inner core is solid. Only P-waves can travel through the inner core because S-waves cannot reach the inner core.

Figure 6.31 shows the path of seismic waves from an earthquake at a point at the top of the diagram. When P-waves enter the outer core they slow down and change direction. Because of the liquid outer core, no waves reach the shadow zone. By investigating seismic waves, scientists have learnt a lot about the Earth's structure.

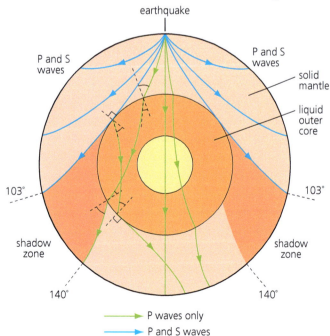

▲ **Figure 6.31**

Test yourself

19 a) What is ultrasound?

b) Ultrasound can be used to examine an unborn baby inside a mother's womb. How do the waves help to build up an image of the baby?

20 Ultrasound is used to investigate a patient's liver. The frequency of the ultrasound is 5 MHz and the speed of the ultrasound is 1600 m/s. Calculate the wavelength of the ultrasound.

21 The ship in Figure 6.30 sends out short pulses of ultrasound every 0.5 s. The frequency of the waves is 50 kHz.

a) Use the information in Figures 6.30a) and 6.30b) to show that the speed of ultrasound in sea water is 1500 m/s.

b) Sketch a graph similar to Figure 6.30b) to show the transmitted and reflected pulses when the ship reaches point A.

c) Calculate the wavelength of the ultrasound waves.

22 Figure 6.32 shows an ultrasonic transmitter and receiver being used to detect cracks in steel rails. The transmitted and reflected pulses are shown on an oscilloscope.

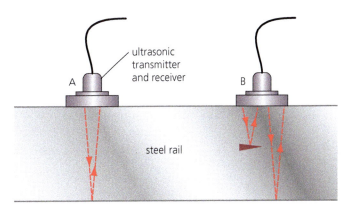

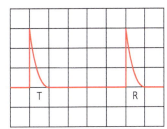 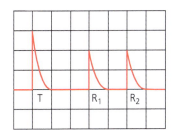

▲ Figure 6.32

a) Why are there two reflected pulses when the detector is in position B? Ultrasound travels at 6000 m/s in steel. The oscilloscope time base measures 10 microseconds per division on the horizontal axis. [1 microsecond = 10^{-6} s]

b) Calculate the thickness of the steel rail.

c) Calculate the depth of the crack in the steel rail.

Show you can...

Give a brief account of how seismic waves provide us with information about the Earth's structure.

Electromagnetic waves

Electromagnetic waves are transverse waves that transfer energy from the source of the waves to an absorber. There are different types of electromagnetic wave which produce different effects and have different uses. The wavelength of electromagnetic waves varies from about 10^{-12} m to over 1 km.

Although electromagnetic waves have a wide range of frequencies and wavelengths, they all have important properties in common.

- They are transverse waves.
- They transfer energy from one place to another.
- They obey the wave equation: $v = f\lambda$
- They travel through a vacuum.
- They travel in a vacuum (space) at a speed of 300 000 000 m/s (3×10^8 m/s).

Electromagnetic waves form a continuous spectrum of wavelengths. Starting with the longest wavelength and going to the shortest, the groups of electromagnetic waves are: radio, microwave, infrared, visible light (red to violet), ultraviolet, X-rays and gamma rays.

Our eyes only detect visible light, so they detect a limited range of electromagnetic waves.

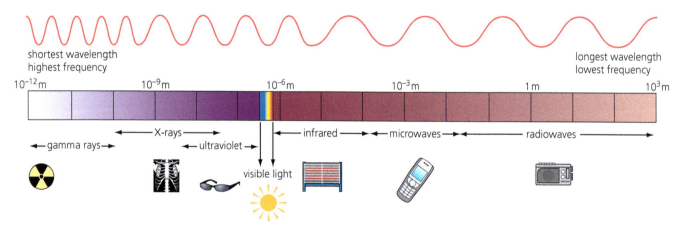

▲ **Figure 6.33** The waves in the electromagnetic spectrum have a range of different uses but also present a number of hazards.

TIP

It is easier to use standard form to solve problems when you are dealing with very large or very small numbers.

Example

A mobile phone uses electromagnetic waves with a frequency of 1.8×10^9 Hz. Calculate the wavelength of the waves.

$$v = f\lambda$$
$$3 \times 10^8 = 1.8 \times 10^9 \times \lambda$$
$$\lambda = \frac{3 \times 10^8}{1.8 \times 10^9}$$
$$\lambda = 0.17\,\text{m}$$

Properties of electromagnetic waves

Different wavelengths of electromagnetic waves are reflected, refracted, absorbed or transmitted differently by different substances and types of surface.

Refraction

When waves travel from one medium to another they usually change direction. This effect is called **refraction** (see Figure 6.34). Different wavelengths refract at different angles. For example, we see rainbows because the different colours of light (which have different wavelengths) are refracted through different angles (Figure 6.35).

Reflection, transmission and absorption

Different wavelengths of visible light behave differently when they are incident on various surfaces. A red shirt appears red because red light is reflected. All other colours (wavelengths) of light are absorbed.

A polished metal surface reflects most electromagnetic waves: wavelengths from radiowaves to ultraviolet are reflected by metal. However, X-rays and gamma rays are able to pass through thin metal.

When a potato is cooked in a microwave oven, it absorbs microwaves. The wavelength of the microwaves is carefully chosen so that water molecules in the food absorb them.

▲ **Figure 6.34** This photograph shows an effect caused as light is refracted (changed in direction) by water in a glass.

▲ **Figure 6.35** Different colours of light refract through different angles.

> ### Test yourself
>
> 23 Give five properties common to all electromagnetic waves.
> 24 Give the meanings of each of the following words, when applied to electromagnetic waves: refraction, reflection, absorption and transmission.
> 25 a) A radio station transmits waves with a frequency of 100 MHz. Calculate the wavelength of the waves.
> b) A radio station transmits waves with a wavelength of 1500 m. Calculate the frequency of the waves.

> ### TIP
>
> Some electromagnetic waves have high frequencies. These can be given in kilohertz kHz (1000 Hz, 10^3 Hz), megahertz MHz (1 000 000 Hz, 10^6 Hz) and gigahertz GHz (1 000 000 000, 10^9 Hz).

Refraction

Figure 6.36 shows how light is refracted as it passes through a rectangular block of glass.

- The angle between the incident ray and the normal is called the angle of incidence.
- The angle between the refracted ray and the normal is called the angle of refraction.

KEY TERM

The **angle of refraction** is the angle between the refracted ray and the normal.

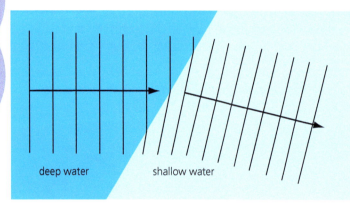

▲ **Figure 6.36** Refraction of light through a glass block.

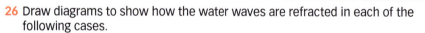

▲ **Figure 6.37** Refraction of water waves.

The light is refracted when it enters and leaves the glass because the speed of light in glass is different to the speed of light in air.

Light changes direction towards the normal when it enters the glass because light travels more slowly in glass than it does in air.

Light changes direction away from the normal when it leaves the glass and enters the air because light travels faster in air than in glass.

Refraction is caused because waves change speed as they pass from one medium to another. All types of waves show refraction. The change of direction of seismic waves (Figure 6.31) as they travel from the mantle to the core is an example of refraction. Seismic waves travel faster in the Earth's mantle than in the core.

Water waves can also be refracted. Figure 6.37 shows what happens when water waves cross from deep water to shallower water at an angle to the edge of the shallower water. In this case the waves:

● slow down
● become shorter in wavelength
● change direction.

Test yourself

26 Draw diagrams to show how the water waves are refracted in each of the following cases.

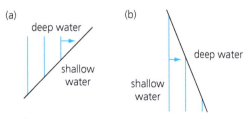

▲ **Figure 6.38**

27 Draw diagrams to show the path of the light rays as they pass through the glass blocks shown in Figure 6.39.

28 This question refers to the path of P-waves shown in Figure 6.31. How can you tell from the diagram that P-waves travel more quickly in the solid mantle than they do in the outer core?

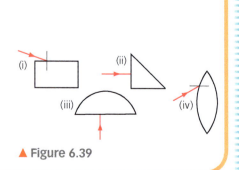

▲ **Figure 6.39**

Required practical 9

Investigating the reflection and refraction of light

Reflection of light

Figure 6.40 shows how the reflection of light from a plane (flat) mirror can be investigated.

Method

1 Using a support, stand a plane mirror upright on a sheet of paper.

2 Draw a line along the back of the mirror.

3 Use a ray box with a single slit to shine a narrow beam of light towards the mirror. This is the incident ray.

4 If you cannot see both the incident and reflected rays on the piece of paper then the mirror is not standing perfectly vertical.

5 Carefully mark the direction of the incident and reflected rays. Do this by drawing two dots at the centre of each ray.

6 Remove the mirror and then join up the two pairs of dots to show the path taken by both rays of light. The two lines that you draw should meet at the line you drew along the back of the mirror.

7 Draw in the normal line where the incident ray hits the mirror.

8 Use a protractor to measure the angle of incidence and the angle of reflection.

9 Record your results in a suitable table.

10 Now repeat the procedure for different angles of incidence.

Analysing the results

1 Plot a graph of the angle of reflection against the angle of incidence.

2 What pattern do the results from your investigation show?

The values that you measured for the angle of incidence and for the angle of reflection may not be 100% accurate. The values are likely to include an uncertainty. This may be ±1°. So an angle of 36° ± 1° would have a true value anywhere between 35° and 37°.

An uncertainty may have been caused by either a random error or a systematic error. Not marking the centre of the rays of light accurately would be a random error. Consistently misusing the protractor to measure the angles would be a systematic error.

Questions

1 How can you tell from your graph if the data includes random errors or a systematic error?

2 How can you reduce the effect of a random error?

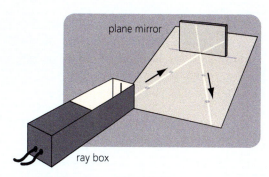

▲ **Figure 6.40**

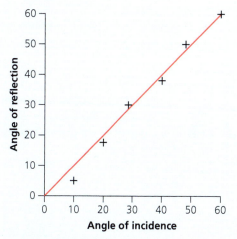

▲ **Figure 6.41** The scattering of the points about the line of best fit shows a random error in the data.

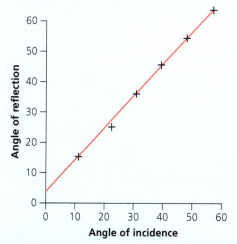

▲ **Figure 6.42** The graph line is higher than expected. It shows that there was a systematic error in measuring the angle of reflection. How much is this error?

Refraction of light

Figure 6.36 (page 196) shows how the refraction of light through a glass block can be investigated.

1 Place the block on a sheet of paper and draw around it.
2 Use a ray box to shine a ray of light into the long side of the block.
3 Mark on the paper where the ray of light enters and leaves the block.
4 Take the block off the paper, mark the path of the light through the block and the normal at the point where the light enters the block. You should now have a diagram that looks like Figure 6.41.
5 Measure the angle of incidence (i) and the angle of refraction (r).
6 Repeat this for other angles of incidence.
7 Record all of your data in a table.

Analysing the results

1 Plot a graph of the angle of incidence against the angle of refraction.
2 How does the angle of refraction compare to the angle of incidence?

Take it further

1 Use the same procedure to investigate the refraction of light by different substances, Find out if the type of substance affects the amount of refraction produced.
2 Use the equipment to show how light can pass through a glass block without being refracted.

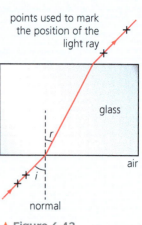

▲ Figure 6.43

○ **Producing electromagnetic waves**

Radiowaves and microwaves

Figure 6.44 shows how radiowaves and microwaves are transmitted and received.

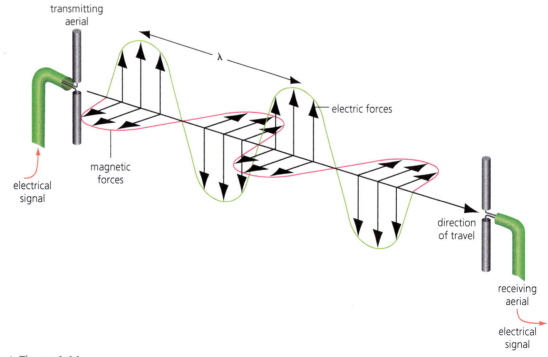

▲ Figure 6.44

- A high frequency alternating current is supplied to the transmitting aerial. This makes electrons oscillate up and down the aerial.
- An electromagnetic wave is emitted. This wave transfers energy in oscillating electric and magnetic fields. This is a transverse wave because the oscillations are at right angles to the direction of energy transfer.
- When the electromagnetic wave reaches the receiving aerial, the electric and magnetic fields cause electrons to oscillate up and down the receiving aerial. This induces a current in an electrical circuit. The alternating current has the same frequency as the radiowave itself.
- Radiowaves also carry the information that we hear and see on radios and televisions.

Changes in atoms

Electromagnetic waves, ranging in wavelength from infrared to X-rays, can be generated or absorbed by changes in atoms.

Gamma rays

Gamma rays are generated when there are changes to the nucleus of an atom. When a gamma ray is emitted, the nuclear energy store of the atom decreases.

⃝ Hazards of electromagnetic waves

Too much exposure to some types of electromagnetic waves can be dangerous. The higher the frequency of the radiation, the more damage it is likely to do to the body. Ultraviolet, X-rays and gamma rays have the highest frequencies.

Ultraviolet, X-rays and gamma rays can all cause mutations to body cells. This can lead to cancer. We are most likely to be exposed to ultraviolet rays when we are in the sun. Ultraviolet waves can cause the skin to age prematurely and increase the risk of skin cancer. We take precautions against ultraviolet radiation by wearing high factor sun cream, sun glasses and by not exposing our skin for a long time. We are less likely to be exposed accidentally to X-rays and gamma rays. However, X-rays and gamma rays are used in medical procedures, where suitable precautions are taken to reduce the risk to our health.

Radiation dose

X-rays and gamma rays are ionising radiations. The damage that these radiations do to human tissue depends on the dose received. The effect of a radiation dose is measured in **sieverts** (Sv). On average we receive a radiation dose of about 2 mSv (2 millisieverts or 2×10^{-3} Sv) per year from background radiation. This dose is thought to be safe, but larger doses can cause harm. You can read more about radiation doses in Chapter 4.

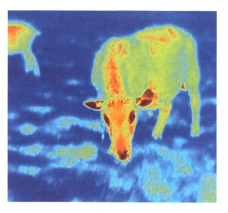

▲ **Figure 6.45** A photograph at night allows us to see the infrared radiation given out by warm bodies.

○ Uses and applications of electromagnetic waves

Electromagnetic waves have many practical applications. Some examples are given on the following pages.

- Radiowaves are used to send radio and television signals from transmitters to our homes. These are called terrestrial signals as they go across the ground. Radiowaves can be transmitted over long distances by reflecting them off the ionosphere (a layer in the Earth's upper atmosphere).
- Microwaves are used for satellite communications because they pass easily through the Earth's atmosphere. Many people now use satellite television. Microwaves are also used in mobile phone networks. These microwaves are transmitted using tall aerial masts.
- Microwaves are used in the home to cook food in microwave ovens. Some wavelengths of microwaves are absorbed by water molecules, and therefore cause heating.
- Infrared waves are used to provide heating from electrical heaters. We cook using infrared radiation in our ovens.
- We can also use infrared cameras to take photographs at night. Infrared has another popular use in remote controls.

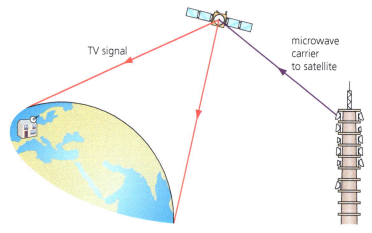

TV signal

microwave carrier to satellite

▲ **Figure 6.46** A satellite dish can transmit a microwave signal that covers a large part of the Earth's surface.

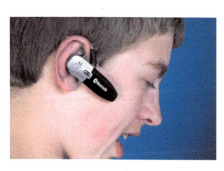

▲ **Figure 6.47** This person is using Bluetooth to link a mobile phone to a hands-free headset.

- We use light waves all the time to see. Light (and infrared) is also used in fibre optic communication systems. Information is coded into signals consisting of light (or infrared) pulses which are then transmitted along fibre optics. Many telephone links now use fibre optics rather than copper cables.
- Lasers that emit high-energy visible light are used for surgery.
- Ultraviolet waves are emitted by hot objects. Some substances can absorb the energy from ultraviolet radiation and then emit the energy as visible light. This is called fluorescence. Some types of energy-efficient lamps work by producing ultraviolet radiation. When electricity passes through a gas in the lamp, reactions occur and ultraviolet radiation is emitted.

The ultraviolet radiation is absorbed by a chemical which covers the inside of the glass. The chemical fluoresces and visible light is emitted. Materials which fluoresce can also have applications in solving crimes (see Figure 6.49).

▲ **Figure 6.48** Ultraviolet radiation can be used to aid the detection of counterfeit money.

▲ **Figure 6.49** This man has been sprayed with 'smartwater'. Each batch of smartwater has a unique chemical signature. If a burglar gets covered with smartwater he can be linked to the scene of a particular crime, when ultraviolet radiation is shone on him.

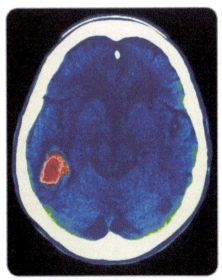

▲ **Figure 6.50** A CT scan of the brain of a patient who has suffered a stroke. The area of the brain affected by the bleeding is shown in red. This is an example of an X-ray image.

- X-rays can penetrate our bodies and therefore provide a valuable way to help doctors diagnose illness or damage to our bodies. In an X-ray photograph, bones, teeth and diseased tissues stand out because they absorb the X-rays.

- Gamma rays are a penetrating radiation which can cause damage to body tissue. Sometimes this is useful. For example, gamma rays are used in radiotherapy to kill cancer cells.

Investigating the emission of infra-red radiation by different surfaces.

Figure 6.51 shows a piece of apparatus called a 'Leslie Cube'. The vertical sides of the cube have different colours or texture. The cube is used to compare the infra-red radiation emitted from each surface.

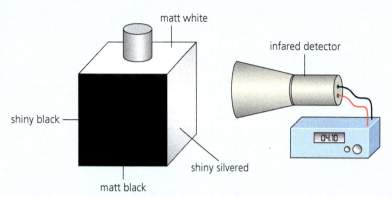

▲ **Figure 6.51** Using a Leslie cube to compare the emission of infra-red radiation from different surfaces.

Method

1 You will be using very hot or boiling water. Write a risk assessment for the investigation.

2 Place the Leslie cube onto a heat proof mat.

3 Fill the cube with very hot water and replace the lid of the cube.

4 Hold an infra-red detector close to one of the sides, wait for the reading to settle and then record the reading in a suitable table.

5 Move the detector and take a reading from each side of the cube. Make sure that the detector is always the same distance from each side of the cube.

If a Leslie cube is not available beakers covered in different materials, for example, aluminium foil and black sugar paper could be used. Fill the beakers with hot water and place a cardboard lid on the beaker. Take the temperature of the water in each beaker after 15 minutes. The cooler the water the faster the surface emits radiation.

Analysing the results

1 Draw a bar chart to show the amount of infra-red emitted against the type of surface.

2 What can you conclude from the results of your investigation?

Questions

1 Why is it important that the detector is held the same distance from each side of the cube?

2 Why is it appropriate to use the readings to draw a bar chart and not a line graph?

Taking it further

Figure 6.52 shows one way of comparing how well two different surfaces absorb infra-red radiation. Before doing this investigation predict which drawing pin will drop off first. Give a reason for your prediction.

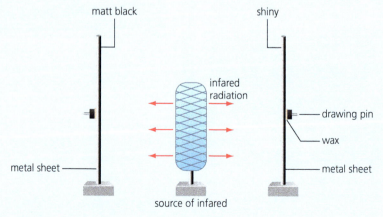

▲ **Figure 6.52** The drawing pin that drops off first is the one attached to the sheet with the surface that absorbs infra-red radiation the fastest.

Show you can...

Show you understand the nature of radiowaves by completing this task.

Describe how an aerial transmits radiowaves, and how an aerial receives radiowaves.

Test yourself

29 Which parts of the electromagnetic spectrum have frequencies higher than ultraviolet waves?
30 Which part of the electromagnetic spectrum:
   a) is used in transmitting terrestrial signals
   b) could cause skin cancer
   c) is used in medical imaging?
31 Choose one part of the electromagnetic spectrum, other than radiowaves, and state one hazard of this wave and one application.

Lenses

○ Refraction in lenses

KEY TERMS

In ray diagrams a **convex lens** is represented by ↕

In ray diagrams a **concave lens** is represented by)(

Many optical instruments use glass lenses to form an image. A lens forms an image by refracting light.

There are two types of lens:
- a **convex** (or converging) **lens**
- a **concave** (or diverging) **lens**.

Figure 6.53 shows how each type of lens refracts light. The rules of refraction apply to both types of lens. When light enters the lens it bends towards the normal. When light leaves the lens it bends away from the normal.

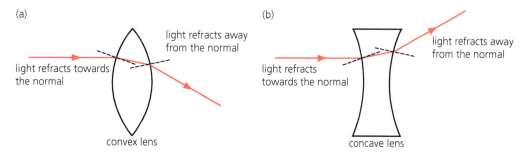

(a) light refracts towards the normal — light refracts away from the normal — convex lens

(b) light refracts towards the normal — light refracts away from the normal — concave lens

▲ **Figure 6.53** The refraction of light by a convex and concave lens.

○ Convex lenses

KEY TERM

The **principal focus** of a convex (converging) lens is the point through which light rays, parallel to the principal axis, pass after refraction.

As a result of refractions, parallel rays of light entering a convex lens converge and meet at a point. This point is the **principal focus** of the lens (Figure 6.54).

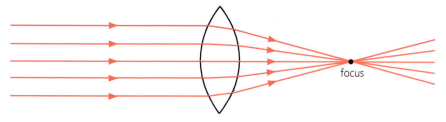

focus

▲ **Figure 6.54**

KEY TERMS ⭐

The **focal length** of the lens is the distance between the lens and the principal focus.

A **real image** is formed when light rays converge to a point. A real image can be projected onto a screen.

TIP ✓

You should know the rules of refraction and be able to predict which way a light ray bends when it crosses a curved surface.

Although you need to understand that light is refracted at both surfaces of a lens, we use a helpful approximation. We make it easier to draw ray diagrams by showing refraction to occur in one place. The refraction in Figure 6.56 is shown to occur at the centre of the lens.

Light can pass through the lens in either direction. So parallel rays coming from the right of the lens converge at a focus an equal distance on the left.

A convex lens can form an image of a distant object onto a screen or a piece of paper. The image is sharpest when the piece of paper or screen is at the principal focus of the lens. The distance between the principal focus and the lens is called the **focal length** of the lens.

When a lens forms an image that we can all see on a screen; it is a **real image**.

▲ Figure 6.55

Ray diagrams

When an object is placed beyond the principal focus, as shown in Figure 6.56, the lens forms a real image of the object.

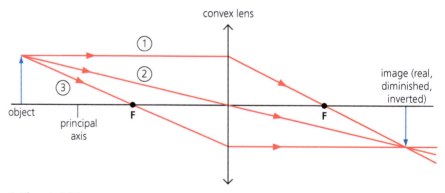

▲ Figure 6.56

Rays come into the lens from all angles, but there are three rays which we can use to locate the image. These rays are chosen because we can predict their path through the lens.

Ray 1 A ray parallel to the principal axis (on the left) passes through the focus on the right.

Ray 2 A ray which passes through the centre of the lens does not change direction.

Ray 3 A ray which passes through the focus of the lens on the left-hand side, emerges from the lens parallel to the principal axis on the right-hand side.

The place where these three rays meet on the right-hand side of the lens locates the top of the image. The bottom of the image lies on the principal axis.

In this example, where the object is a long way from the lens, the image is:

- real
- inverted
- diminished in size (smaller).

A real image is formed when rays meet at a point, and the image can be seen on a screen.

To find the position of an image you only need to draw two of the rays, not all three. So from now on we will just use rays 1 and 2.

Figure 6.57 shows the size and position of the image when the object is just outside the focal length of the lens.

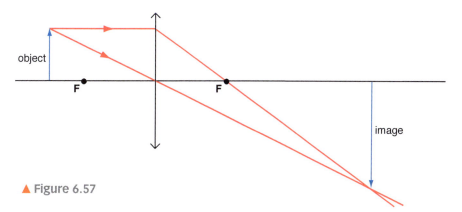

▲ **Figure 6.57**

The magnification produced by a lens is given by the equation:

$$\text{magnification} = \frac{\text{image height}}{\text{object height}}$$

Magnification is a ratio of two heights and so has no units.

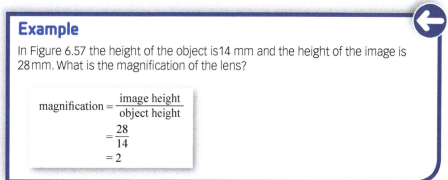

Example

In Figure 6.57 the height of the object is 14 mm and the height of the image is 28 mm. What is the magnification of the lens?

$$\text{magnification} = \frac{\text{image height}}{\text{object height}}$$
$$= \frac{28}{14}$$
$$= 2$$

A magnifying glass

A single convex lens can be used as a magnifying glass. The most powerful magnifying glasses use thick lenses.

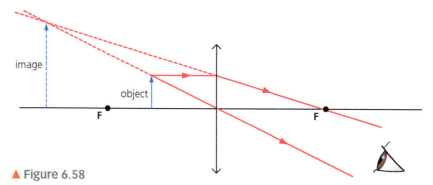

▲ Figure 6.58

When an object is placed inside the focal length of a lens, the rays of light that pass through the lens do not meet. When these rays are viewed by the eye, they appear to come from behind the lens. This image is only seen by the person looking through the lens.

The image is:

● virtual
● upright (the right way up)
● magnified.

○ Concave lenses

As a result of refraction, parallel rays of light entering a concave (or diverging) lens spread out as they pass through the lens (Figure 6.59). The rays look as though they spread out or diverge from a single point. This point is the principal focus of the lens – but because the light only seems to diverge from this point, it is a **virtual focus**.

Images and ray diagrams

Ray diagrams can be drawn for concave lenses in the same way as for convex lenses. A ray of light incident on the centre of the lens does not change direction. A ray of light that is parallel to the principal axis is refracted so that it seems to have come from a virtual focus.

A virtual image is formed where the two rays seem to cross. It does not matter where the object is, inside or outside the focal length of the lens, the nature of the image is always the same:

● virtual
● upright
● diminished.

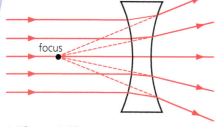

▲ Figure 6.59

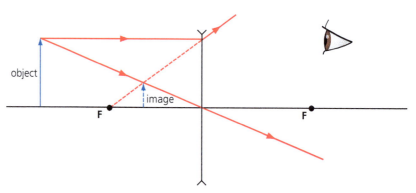

▲ Figure 6.60

Test yourself

32 Copy and complete the ray diagrams drawn in Figure 6.61. 'F' marks the principal focus of the lens.

(a)
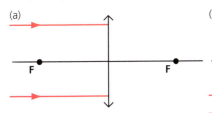

(b)
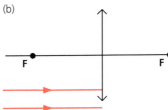

(c)
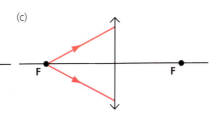

▲ Figure 6.61

33 a) A concave lens has a 'virtual' principal focus. What does this mean?

 b) Why is a concave lens no good as a magnifying glass?

34 a) What is meant by **i)** a real image, **ii)** a virtual image?

 b) Which type of lens always produces a virtual image?

35 Look at the lenses in Figure 6.62. Which are convex lenses?

36 Calculate the magnification of the images in Figures 6.56 and 6.58.

37 a) Copy and complete the diagram in Figure 6.63 to show how the lens forms an image.

 b) State whether the image is
  i) real or virtual
  ii) magnified or diminished
  iii) upright or inverted.

38 A convex lens produces a magnification of ×1.5 when a real image is formed at a distance of 15 cm from the lens.

 a) Draw a scale diagram to show how the image is formed.

 b) Use your diagram to calculate the focal length of the lens.

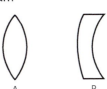

A   B   C   D

▲ Figure 6.62

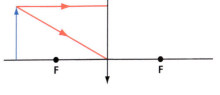

▲ Figure 6.63

Show you can...

Complete this task to show you understand how to draw ray diagrams to locate the position of virtual images.

Draw a ray diagram to show how a convex lens can be used as a magnifying glass.

Visible light

TIP ✓
You can remember the order of the colours using the mnemonic: Richard of York gave battle in vain.

A **prism** can be used in a laboratory to split white light into different colours. Each colour travels at a slightly different speed and is refracted through a different angle as shown in Figure 6.64.

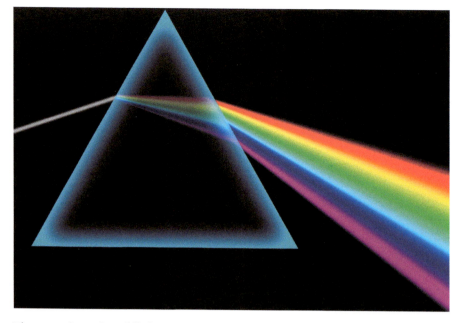

◀ **Figure 6.64** Isaac Newton identified seven colours of the rainbow: red, orange, yellow, green, blue, indigo, violet.

The wavelengths of light range from about 400×10^{-9} m (or 400 nm, 400 nanometres) to about 700×10^{-9} m (Figure 6.65).

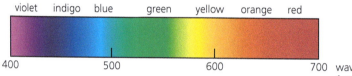

▲ **Figure 6.65** The spectrum of visible light from violet to red.

○ Seeing colours by reflection

You have already met the idea of reflection off a smooth polished surface such as a mirror. Light is reflected in a single direction and we can see an image. This is called **specular reflection** (see Figure 6.66).

When light is incident on a rough surface, light is reflected (or scattered) in all directions. This is called **diffuse reflection**. A painted wall scatters light in all directions. We do not see an image, we just see the colour of the wall (see Figure 6.67).

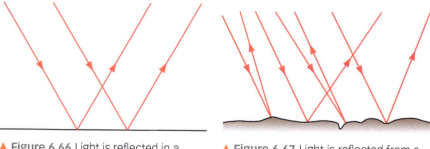

▲ **Figure 6.66** Light is reflected in a single direction from a smooth surface.

▲ **Figure 6.67** Light is reflected from a rough surface in all directions.

The colour of an **opaque** object is determined by which wavelengths of light are more strongly reflected. Any wavelengths which are not reflected are absorbed.

KEY TERMS ⭐

Specular reflection occurs when light is reflected off a smooth surface in a single direction.

Diffuse reflection occurs when light is reflected at different angles off a rough surface.

KEY TERM ⭐

An **opaque** object does not allow light to pass through.

For example:

- In Figure 6.68a) the object looks blue because only blue light is reflected. All other colours are absorbed.
- In Figure 6.68b) the object looks red because only red light is reflected and all other colours are absorbed.
- In Figure 6.68c) the object looks green because green light is strongly reflected. All the other colours are absorbed.

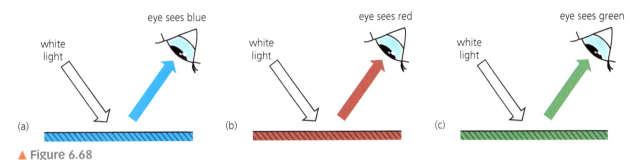

▲ Figure 6.68

An object which reflects all wavelengths equally looks white. An object which absorbs all wavelengths completely looks black.

Some objects transmit light; these are either **transparent** or **translucent**.

Coloured filters

A **filter** allows only a particular small range of wavelengths to pass through. For example, a green filter allows only green light to pass through it and it absorbs all other colours of light.

In Figure 6.69 the red filter allows red light to pass through but absorbs all the other colours. When the red light reaches a green filter, no light passes through, because the filter absorbs the red light.

If you look at a blue object through a blue filter it looks blue. However, if you look at a red object through a blue filter the object looks black.

KEY TERMS

A **transparent** object allows us to see clearly through it. Glass is transparent.

A **translucent** object allows light to pass through, but we cannot see objects through it clearly. Some plastics and frosted glass are translucent.

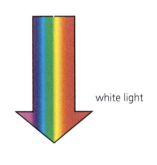

white light

red filter

red light

green filter

no light

▲ Figure 6.69

Test yourself

39 What colour will a blue car appear if you look at it through

  a) a blue filter   **b)** a red filter   **c)** a green filter?

40 A girl goes to a disco wearing red shoes, green trousers, a blue shirt and a black and white cap. Copy and complete the table to show what colour her clothing looks in various colours of disco lights.

Colour of light	Shoes	Trousers	Shirt	Cap
White				
Red				
Blue				
Green				

41 Explain how the reflection and absorption of different wavelengths of light cause the colours of the objects we see.

42 Why do different colours of light refract through different angles when white light enters a prism?

43 Light travels through air at a speed of 3×10^8 m/s.

  a) Calculate the frequency of red light with a wavelength of 6.5×10^{-7} m.

  b) Calculate the wavelength of yellow light with a frequency of 5.2×10^{14} Hz.

Black body radiation

If you put your hand close to your face, both your hand and your face feel a little warmer. This is because your body emits and absorbs infrared radiation. This radiation gives us the sensation of warmth. All objects, whatever their temperature, emit and absorb infrared radiation.

We also know from experience that an electric fire emits infrared radiation. When you first turn an electric fire on you can feel the radiation reaching your hand. After a while you have to take your hand away because the fire gets hotter and begins to glow red. The hotter the fire, the more infrared radiation is emitted per second.

The colour of a hot object also tells us its temperature. When an object is at a temperature of about 500 °C, it begins to glow red-hot. At a temperature of 1000 °C the colour of a hot object is a yellow orange. At higher temperatures objects begin to glow white hot.

- All objects emit and absorb infrared radiation no matter what their temperature is.
- The hotter the object, the more radiation is emitted per second.
- As the temperature of an object increases, more energy is emitted as visible light.
- The shorter the wavelength of visible light emitted from a hot object, the higher the temperature.

▲ **Figure 6.70** A blacksmith can judge the temperature of the steel by looking at the colour of the radiation.

A perfect black body

A perfect black body is one that absorbs all the radiation that is incident on it. A perfect black body does not reflect or transmit any radiation through it. Since a good absorber is also a good emitter, a black body is also the best possible emitter.

In practice there is no such thing as the perfect black body, but objects coated with substances such as lamp black absorb about 98% of the radiation incident on them.

○ **Radiation and temperature change**

Figure 6.71 shows two objects which have been put into an oven. Each has a temperature of 180 °C. The two objects absorb different amounts of radiation.

Object X is a perfect black body.

- X absorbs all the radiation falling on it per second.
- X re-emits the same amount of radiation per second.

Object Y reflects half the energy that falls on it.

- Y absorbs half the radiation and reflects half the radiation falling per second.
- Y re-emits the same amount of radiation that it absorbs each second.

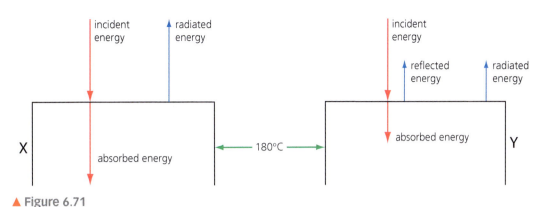

▲ Figure 6.71

When an object is at a constant temperature, it is absorbing radiation at the same rate as it is emitting radiation.

If an object absorbs more radiation than it emits, the temperature of the object rises. For example:

- A cold room is warmed up by an electric fire.
- The Earth's surface warms up each day as the Sun shines on it.
- Meat is heated by radiation inside an oven.

If an object emits more radiation than it receives, the object cools down. For example:

- A hot piece of coal cools down when taken out of a fire.
- At night the Earth cools down.

KEY TERM

The **intensity of radiation** is the power of the radiation incident per square metre.

The Earth and radiation

The temperature of the Earth depends on many factors including the rates of absorption and emission of radiation, and also the amount of radiation reflected into space.

Figure 6.72 helps to explain two well-known facts about our Earth.

- It is much hotter at the equator than it is at the poles of the Earth. At the equator the **intensity of radiation** is high. At the poles the intensity of radiation is much lower, because sunlight strikes the Earth's surface at an angle.

- At night time the part of the Earth's surface facing away from the Sun cools. There is no radiation arriving from the Sun and the Earth's surface radiates energy into space.

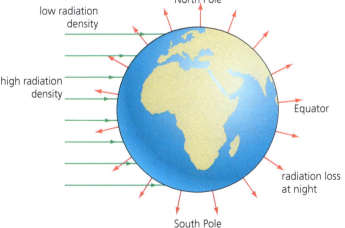

▲ Figure 6.72

Figures 6.73 and 6.74 explain how clouds can affect our weather.

At position A in Figure 6.73 the temperature is hotter than it is at B, because the clouds above B reflect some of the radiation. This means that B warms up more slowly than A as the Sun rises.

In Figure 6.74 it is night time. At night the Earth cools down due to radiation loss. Now cloud cover keeps us warm. At C the Earth radiates energy into space, but at position D the clouds absorb and reflect some of the radiation. D remains warmer than C, as the clouds reflect some radiation back to the Earth's surface.

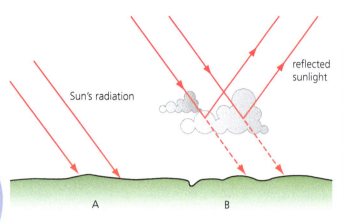

▲ Figure 6.73

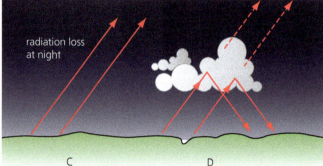

▲ Figure 6.74

Test yourself

44 Copy and complete these sentences.
  All bodies at all temperatures emit _____ radiation.
  At higher temperatures hot bodies emit _____ light.
45 Explain how a blacksmith can tell the approximate temperature of a metal without using a thermometer.
46 What is meant by a perfect black body?
47 Draw diagrams to explain these facts about the Earth.
  a) The Earth is hotter at the equator than at the poles.
  b) The Earth cools down at night.

Show you can...

Show you understand about infrared radiation by completing the task.

Explain why clouds keep the Earth cooler in the day but warmer at night.

Chapter review questions

1 a) Explain the difference between a longitudinal and a transverse wave.

   b) Give two examples of each type of wave.

2 Name a type of electromagnetic wave that:

   a) is emitted by hot objects

   b) is used to communicate with satellites

   c) causes a sun tan.

3 A tuning fork vibrates at a frequency of 512 Hz. The speed of sound is 330 m/s.

   a) Calculate the period of the vibration of the tuning fork.

   b) Calculate the wavelength of the sound waves produced by the tuning fork.

4 A ship is close to a cliff and sounds its fog horn. An echo is heard.

   a) What causes the echo?

   Sound travels at 330 m/s. The echo is heard after 4 seconds.

   b) Calculate the distance from the ship to the cliff.

5 End A of the rope in Figure 6.75 is shaken up and down at a rate of five oscillations every 2 seconds.

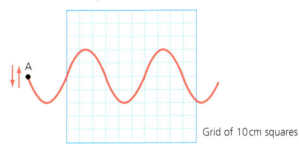

Grid of 10 cm squares

▲ **Figure 6.75**

   a) Determine the amplitude of the wave motion (maximum displacement to one side).

   b) Determine the wavelength.

   c) Calculate the time period.

   d) Calculate the frequency.

   e) How far do the waves travel during one time period?

   f) Calculate the wave speed.

6 a) Copy each of the diagrams in Figure 6.76 and draw ray diagrams to show how each lens forms an image of the object.

   b) Give a description of each type of image.

   c) Calculate the magnification of each image.

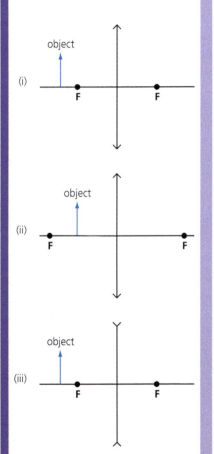

▲ Figure 6.76

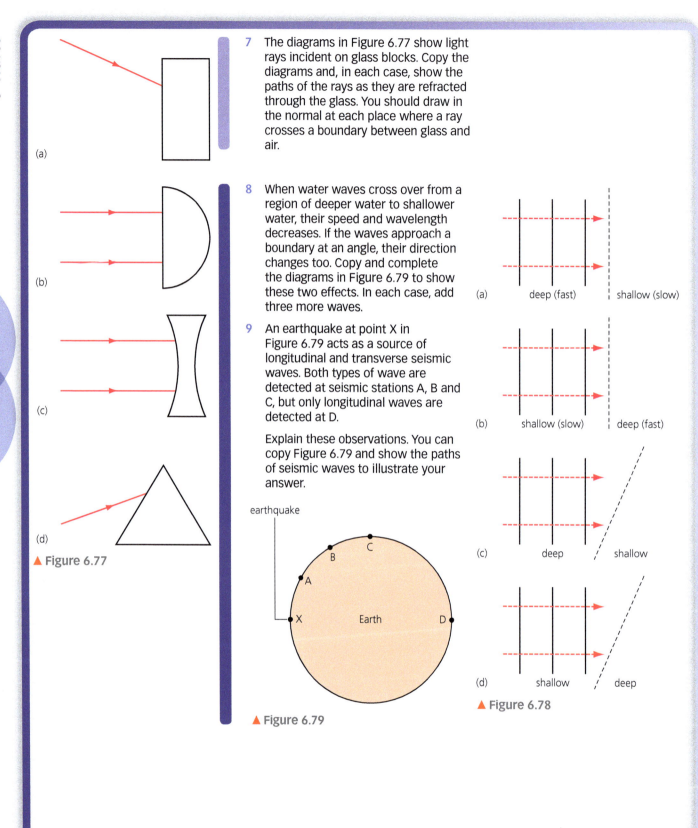

(a)

(b)

(c)

(d)

▲ Figure 6.77

7 The diagrams in Figure 6.77 show light rays incident on glass blocks. Copy the diagrams and, in each case, show the paths of the rays as they are refracted through the glass. You should draw in the normal at each place where a ray crosses a boundary between glass and air.

8 When water waves cross over from a region of deeper water to shallower water, their speed and wavelength decreases. If the waves approach a boundary at an angle, their direction changes too. Copy and complete the diagrams in Figure 6.79 to show these two effects. In each case, add three more waves.

9 An earthquake at point X in Figure 6.79 acts as a source of longitudinal and transverse seismic waves. Both types of wave are detected at seismic stations A, B and C, but only longitudinal waves are detected at D.

Explain these observations. You can copy Figure 6.79 and show the paths of seismic waves to illustrate your answer.

earthquake

Earth

▲ Figure 6.79

(a) deep (fast)   shallow (slow)

(b) shallow (slow)   deep (fast)

(c) deep   shallow

(d) shallow   deep

▲ Figure 6.78

Practice questions

1 a) Use the correct words from the box to complete the sentence.

absorbs	ionises	reflects	transmits

When X-rays enter the human body, soft tissue .... X-rays and bone .... X-rays. [2 marks]

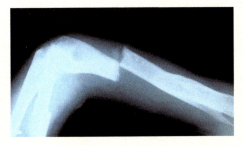

▲ Figure 6.80

b) Which one of the following is another use of X-rays?

- scanning unborn babies
- providing a sun tan
- killing cancer cells [1 mark]

c) The table shows the dose received by the human body when different parts are X-rayed.

Part of body X-rayed	Dose received by the human body in millisieverts
Chest	3
Skull	15
Back	120

How many chest X-rays would give a patient the same dose as one back X-ray? [2 marks]

2 A person looks through a magnifying glass at an object. The ray diagram shows how the lens produces an image.

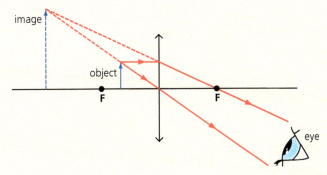

▲ Figure 6.81

a) Which two words from the list describe the nature of the image?

Write down the two correct answers. [2 marks]

magnified	diminished	inverted	real	virtual

b) Use the information from the ray diagram to calculate the magnification of the image. [2 marks]

3 Light from an object forms an image in a plane mirror.

a) Copy out the **two** statements that are correct. [2 marks]

- The image in a plane mirror is virtual.
- Light from the object passed through the image in a plane mirror.
- Light waves are longitudinal.
- The angle of incidence equals the angle of reflection.
- The incident ray is always at right angles to the reflected ray.

b) i) Copy Figure 6.82 and use words from the box to label the numbers on the diagram. [2 marks]

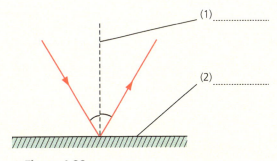

▲ Figure 6.82

mirror	normal	ray	reflection

ii) Use the letter 'r' to label the angle of reflection. [1 mark]

4 A teacher and two students are measuring the speed of sound.

150m

▲ Figure 6.83

The teacher makes a sound by hitting two cymbals together. Each student starts a stopwatch when they see the teacher hit the cymbals. They each stop their stopwatch when they hear the sound.

a) Describe how a sound wave moves through the air. [3 marks]

b) The teacher repeats the experiment and the students record the readings on their stopwatch in a table.

Student	Time in seconds
1	0.44, 0.46, 0.44, 0.48, 0.43
2	0.5, 0.6, 0.4, 0.4, 0.6

    i) State the resolution of the first student's readings. [1 mark]

    ii) State the equation linking speed, distance travelled and time taken. [1 mark]

c) The teacher was standing 150 m from the students. Use the experimental data recorded by each student to calculate:

    i) the average time recorded by each student [2 marks]

    ii) the speed of sound calculated by each student. [2 marks]

Write each answer to an appropriate number of significant figures.

5 The diagram represents the electromagnetic spectrum.

radiowaves	microwaves	infrared	A	ultraviolet	B	gamma rays

a) Name the two parts of the spectrum labelled A and B. [2 marks]

b) Which electromagnetic radiation is used for night vision equipment? [1 mark]

c) Which electromagnetic radiation is used for transmitting signals to a satellite? [1 mark]

d) Which type of electromagnetic wave has the greatest frequency? [1 mark]

e) Exposure to excessive electromagnetic radiation can be harmful to the human body. For two named types of radiation, describe:

    i) a harmful effect for each type [2 marks]

    ii) how the risks of exposure can be reduced. [2 marks]

6 a) Explain why a red flower looks red when seen in white light. [2 marks]

b) What colour does the red flower appear to be when seen in

    i) red light

    ii) green light? [2 marks]

7 Figure 6.84 shows a wave on the sea.

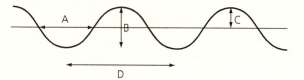

▲ **Figure 6.84**

a) **i)** Which letter shows the wavelength of the wave? [1 mark]

    ii) Which letter shows the amplitude of the wave? [1 mark]

b) What type of wave is shown in the diagram? Describe how energy is transferred by the water. [2 marks]

c) A man watches some waves pass his boat. One wave passes him every 4 s.
Calculate the frequency of the waves. [2 marks]

8 a) What is ultrasound? [1 mark]

b) In Figure 6.85 ultrasound is used to measure the depth of the seabed. A pulse of ultrasound is sent out from the bottom of the ship. It takes 1.2 seconds for the emitted sound to be received back at the ship.

Calculate the depth of the water.

Speed of ultrasound in water = 1500 m/s [3 marks]

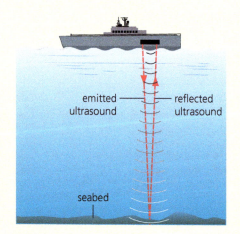

▲ **Figure 6.85**

c) Ultrasound can be used in medicine for scanning.

State one medical use of ultrasound scanning. [1 mark]

9 a) A lens is used to project an image of an object.

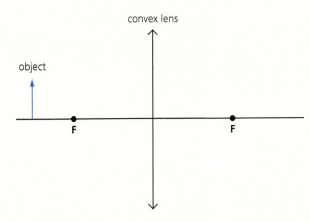

▲ Figure 6.86

Copy and complete the ray diagram to show how the lens produced an image of the object. **[3 marks]**

b) State three words to describe the nature of the image. **[3 marks]**

c) Use your diagram to calculate the magnification of the lens. **[2 marks]**

10 a) A person looks at an object through a concave lens.

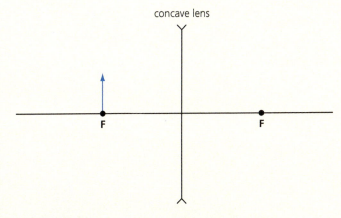

▲ Figure 6.87

Copy and complete the ray diagram to show how the lens produces an image of the object. **[4 marks]**

b) This lens produces a virtual image. Explain the difference between a virtual and a real image. **[2 marks]**

11 Earthquakes produce two types of seismic wave, an S-wave, which is transverse, and a P-wave which is a longitudinal wave.

a) Explain the difference between a longitudinal and a transverse wave. **[2 marks]**

Figure 6.88 shows the two types of wave reaching the boundary between the Earth's mantle and the Earth's outer core. The mantle is solid, and the outer core is liquid.

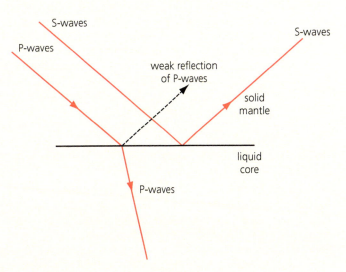

▲ Figure 6.88

b) Explain why all the S-waves are reflected from the boundary. **[2 marks]**

c) What can you deduce about the speed of the P-waves in the mantle, compared with the speed of the P-waves in the outer core? Give a reason for your answer. **[2 marks]**

Working scientifically

Communicating scientific results and developments

Reports of events or scientific developments in the newspapers, on television and on the internet may often be oversimplified, inaccurate, biased or simply confusing. Following the nuclear accident at Fukushima in Japan, newspaper reports about the dangers of radioactivity caused a lot of uncertainty and concern for many people living in Japan. Recent reports in the media suggest a link between the increase in the numbers of children having cancer of the thyroid and the nuclear accident. Health scientists say this link has not been substantiated.

Peer review

One of the problems with media reports is that they are not subjected to peer review.

Peer review is a process scientists use to try and make sure that scientific results, developments and reports are correct. The work of one group of scientists will be studied and checked by another group of scientific experts. They will look at the evidence and the way in which the evidence was obtained. Peer review is a way of having someone who understands the subject checking over another person's work. Peer review usually, but not always, takes place before scientific work is published and made available to a wider audience.

> **KEY TERM**
> **Peer review** is a process by which scientists check each other's work.

Particles measured travelling faster than the speed of light

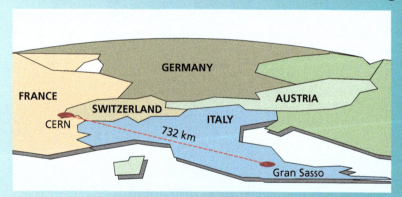

▲ **Figure 6.89** The beam of neutrinos travelled from Switzerland to Italy.

In 2011, a group of physicists published the results of an experiment called OPERA. The experiment had been carried out over three years. It involved timing a beam of particles, called neutrinos, travelling 730 km from the CERN laboratory in Switzerland to the Gran Sasso laboratory in Italy. The results were amazing; they suggested that the particles could travel faster than the speed of light!

Having measured over 16 000 events, the physicists decided to publish their results for checking and review by other groups of physicists. In 2012 a mistake was found in the experiment which had caused an error in the original results. The time taken by the neutrinos to travel the 730 km was in fact always 60 nanoseconds (60×10^{-9} s) more than the time recorded in the experiment.

1 Calculate how long it would take a neutrino travelling at the speed of light to go from CERN to Gran Sasso.

2 What type of error did the physicists in the OPERA experiment make?

3 Why was it important that the physicists published the results of the OPERA experiment?

Fact or opinion?

Headlines in newspapers are designed to grab your attention, to make you want to read the rest of the article. Mobile phones and phone masts are often in the headlines.

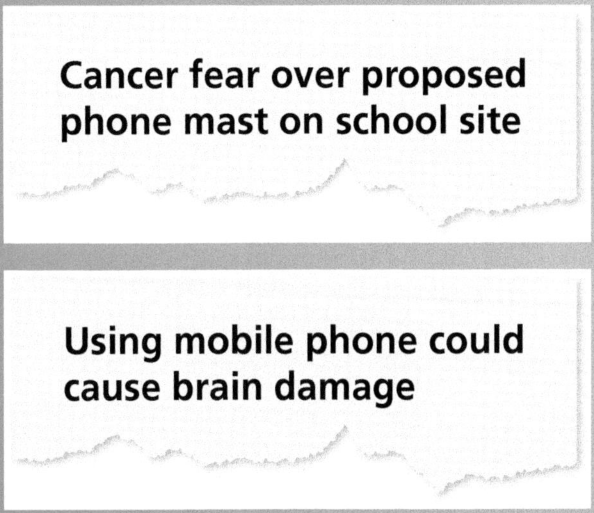

Cancer fear over proposed phone mast on school site

Using mobile phone could cause brain damage

▲ Figure 6.90

4 Do the newspaper headlines give facts or opinions?

There has been a lot of scientific research into the potential health effects of using a mobile phone. Some studies have suggested a link between mobile phone use and health problems such as tiredness, headaches and brain tumours. A study of 750 people in Sweden led to the suggestion that using a mobile phone over a period of at least 10 years increases the risk of a tumour on the nerve between the ear and the brain by four times. However, the evidence from these and other studies is not conclusive. What the evidence does suggest is a need for more research and for continued peer review of new evidence.

5 What do scientists mean by conclusive evidence?

6 Do you think that the number of people studied in Sweden was enough to give firm evidence of a link between mobile phone use and developing a tumour? Give reasons for your answer.

7 Magnetism and electromagnetism

At home you might use a magnet for a purpose as simple as attaching your shopping list to the fridge door. The photograph opposite shows a completely different application of magnetism. Here you can see the central view of a particle detector at the Large Hadron Collider in the European Centre for Nuclear Research (CERN). The particle detector is an arrangement of eight magnetic coils (each weighing 100 tonnes), which are used to deflect high-energy particles into the central detector. The coils are cooled to a temperature of −269 °C at which temperature the coils are superconducting – this means that the coils have no electrical resistance. A current of about 21 000 A is used to produce very strong magnetic fields.

Specification coverage

This chapter covers specification points: **4.7.1** Permanent and induced magnetism, magnetic forces and fields, **4.7.2** The motor effect and **4.7.3** Induced potential, transformers and the National Grid.

Prior knowledge

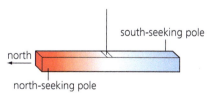

▲ Figure 7.1

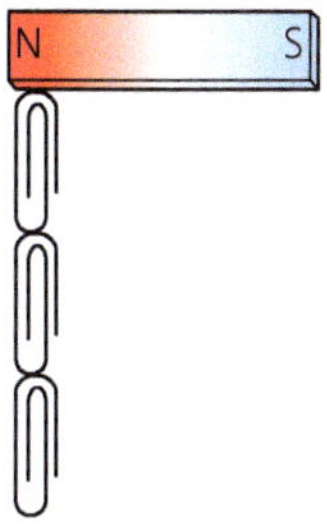

▲ Figure 7.2

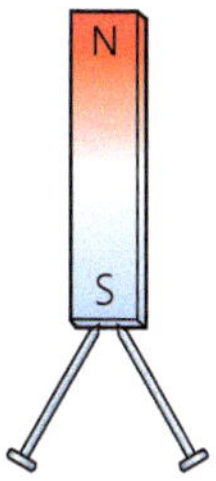

▲ Figure 7.3

Previously you could have learned:

> Some materials are magnetic. The most common magnetic materials are iron and steel.

> Magnets have two poles, north and south.

> Magnets attract magnetic materials at a distance. Magnetism is a non-contact force.

> Two like poles repel each other: a north pole repels a north pole; a south pole repels a south pole.

> Two unlike poles attract: a north pole attracts a south pole.

Test yourself on prior knowledge

1 Copy Figure 7.1 and add arrows to show the direction of the magnetic forces acting on each magnet.

2 In Figure 7.2, three steel paper clips are attracted to a magnet. Copy the sentences below and fill in the gaps to explain why this happens.
   • The magnetic field of the magnet . . . . the paper clips.
   • The top of each clip becomes a . . . . magnetic pole and the bottom of each paper clip becomes a . . . . magnetic pole. Each clip attracts the one below. The size of the magnetic attraction is greater than the . . . . of each clip.

3 Explain why the steel pins repel each other in Figure 7.3.

Permanent and induced magnetism, magnetic forces and fields

○ Poles

KEY TERMS ⭐

Magnetic materials are attracted by a magnet.

A north-seeking pole is the end of the magnet that points north.

A south-seeking pole is the end of the magnet that points south.

Some metals, for example iron, steel, cobalt and nickel, are **magnetic**. A magnet will attract them. If you drop some steel pins on the floor, you can pick them up using a magnet. A magnetic force is an example of a non-contact force, which acts over a distance.

In Figure 7.4, you can see a bar magnet which is hanging from a fine thread. When it is left for a while, one end always points north. This end of the magnet is called the **north-seeking pole**. The other end of the magnet is the **south-seeking pole**. We usually refer to these poles as the north and south poles of the magnet.

The magnetic forces on steel pins, iron filings and other magnetic objects are always greatest when they are near the poles of a magnet. Every magnet has two poles which are equally strong.

When you hold two magnets close together you find that two north poles (or two south poles) repel each other, but a south pole attracts a north pole (Figure 7.5).

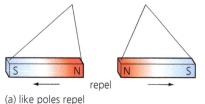

▲ **Figure 7.4** This suspended bar magnet has a north pole and a south pole.

(a) like poles repel

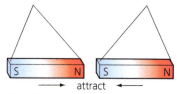

(b) unlike poles attract

▲ **Figure 7.5** Two like poles will repel, while unlike poles attract.

Plotting fields

Method

Use a bar magnet and a small plotting compass as shown in Figure 7.6.

Place the compass close to the north pole. Put a small mark at the end of the compass to show the direction of the field. Move the compass to see how the field changes direction.

Make a sketch of the shape of the magnetic field. Note that when you make a sketch, your field lines should not cross because the field can only point in one direction.

Magnetic fields

There is a magnetic field in the area around a magnet. In this area, a force acts on a magnetic object or another magnet. If the field is strong, the force is big. If the field is weak, the force is small.

The direction of a magnetic field can be found by using a small plotting compass. The compass needle always points along the direction of the field. Figure 7.6 shows how you can investigate the magnetic field near to a bar magnet using a compass.

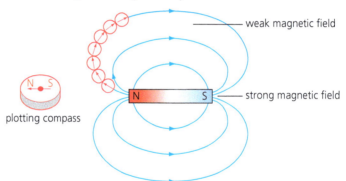

▲ **Figure 7.6** Investigating the magnetic field near to a bar magnet using a compass.

We use magnetic field lines to represent a magnetic field. Magnetic field lines always start at a north pole and finish at a south pole. When the field lines are close together, the field is strong. The further apart the lines are, the weaker the field is. A magnetic field is strongest at the poles of a magnet and gets weaker as the distance from the magnet increases.

Magnetic field lines are not real, but are a useful model to help us to understand magnetic fields.

Compass

Figure 7.7 shows a compass that you can use to find north when you are walking. A compass contains a small bar magnet that can rotate. When a compass is held at rest in your hand, the needle always settles along a north-south direction. This behaviour provides evidence that the Earth has a magnetic field. The pole that points towards north is the a north-seeking pole. Since unlike poles attract, this tells us that at the magnetic north pole, there is a south-seeking pole. Figure 7.8 shows the shape of the Earth's magnetic field.

▲ **Figure 7.7** A compass, which you can use to find north when you are walking.

North Pole

magnetic north

S N

N S

N

◀ **Figure 7.8** The shape of the Earth's magnetic field.

South Pole

KEY TERMS ⭐

A permanent magnet produces its own magnetic field. It always has a north pole and a south pole.

An induced magnet becomes magnetic when it is placed in a magnetic field.

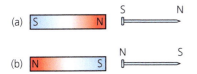

(a) S N   S N

(b) N S   N S

▲ **Figure 7.9** The nail becomes magnetised by the magnet's field. The nail is always attracted to the magnet.

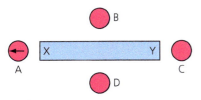

▲ **Figure 7.10**

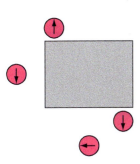

▲ **Figure 7.11**

Permanent and induced magnets

Some magnets are **permanent magnets**. Permanent magnets produce their own magnetic field. They always have a north and south pole. You can check to see if a magnet is a permanent magnet by placing it near to another magnet that you know is permanent. If both magnets are permanent, then you will be able to see that they can repel each other, as well as attract.

An **induced magnet** is a material that becomes magnetic when it is placed in a magnetic field. Induced magnets are temporary magnets. An induced magnet is always attracted towards a permanent magnet. This is because the induced magnet is magnetised in the direction of the permanent magnet's field. When the induced magnet is taken away from the permanent magnetic field, it will lose all (or most) of its magnetism quickly.

Test yourself

1 Which of the following items will a bar magnet pick up?
Brass screw   Steel pin   Iron nail   Aluminium can
2 a) Figure 7.10 shows a bar magnet surrounded by four plotting compasses. Copy the diagram. Mark in the direction of the compass needle for the positions B, C and D.
   b) Which is the north pole of the magnet, X or Y? Give the reason for your answer.
3 Two bar magnets have been hidden in a box. Use the information in Figure 7.11 to suggest how they are arranged inside the box.
4 a) Explain what is meant by a permanent magnet.
   b) Explain what is meant by an induced magnet.

Show you can...

Complete this task to show you understand the nature of magnets.

Plan an experiment to show the difference between permanent and induced magnets.

The motor effect

bird's eye view

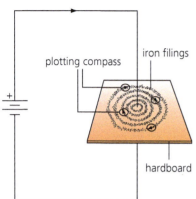

▲ **Figure 7.12** This experiment shows there is a magnetic field around a current-carrying wire.

KEY WORD ⭐

Right-hand grip rule A way to work out the direction of the magnetic field in a current-carrying wire if you know the direction of the current.

⭕ The magnetic field near a straight wire

An electric current in a conducting wire produces a magnetic field around the wire.

In Figure 7.12, a long straight wire carrying an electric current is placed vertically so that it passes through a horizontal piece of hardboard. Iron filings have been sprinkled onto the board to show the shape of the field. Here is a summary of the important points of the experiment.

- When the current is small, the magnetic field is too weak to notice. However, when a large current is used, the iron filings show a circular magnetic field pattern (see Figure 7.13).
- The magnetic field gets weaker further away from the wire.
- The direction of the magnetic field can be found using a compass. If the current direction is reversed, the direction of the magnetic field is reversed.

Figure 7.14 shows the pattern of magnetic field lines surrounding a wire. When the current flows into the paper (shown ⊗) the field lines point in a clockwise direction around the wire. When the current flows out of the paper (shown ⊙) the field lines point anti-clockwise.

The right-hand grip rule will help you to remember this (Figure 7.15). Put the thumb of your right hand along a wire in the direction of the current. Your fingers will point in the direction of the magnetic field.

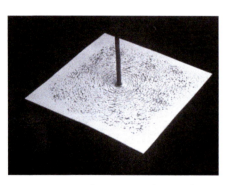

▲ **Figure 7.13** The circular shape of the magnetic field is shown by the pattern of iron filings.

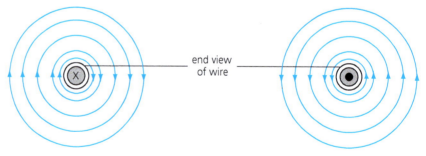

end view of wire

▲ **Figure 7.14** The direction of the magnetic field depends on the direction of the current.

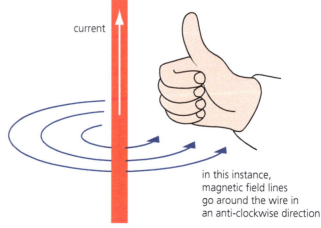

current

in this instance, magnetic field lines go around the wire in an anti-clockwise direction

▶ **Figure 7.15** The right-hand grip rule.

○ The magnetic field of a solenoid

KEY WORD

Solenoid A solenoid is a long coil of wire, as shown in Figure 7.16.

Figure 7.16 shows the magnetic field that is produced by a current flowing through a long coil of wire or **solenoid**. The magnetic field from each loop of wire adds on to the next. The result is a magnetic field that is like that of a long bar magnet.

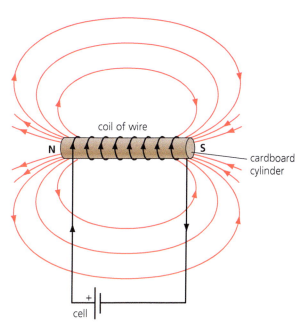

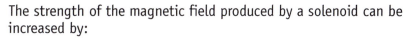

▲ **Figure 7.16** The magnetic field around a solenoid is similar in shape to that of a bar magnet.

The strength of the magnetic field produced by a solenoid can be increased by:

● using a larger current
● using more turns of wire
● putting the turns closer together
● putting an iron core into the middle of the solenoid.

○ **Electromagnets**

▲ Figure 7.17 An electromagnet in action.

Figure 7.17 shows a practical laboratory electromagnet. It is made into a strong magnet by two coils with many turns of wire which increase the magnetising effect of the current. When the current is switched off, the magnet loses its magnetism and so the iron filings fall off.

Relays

A car starter motor needs a very large current of about 100 A to make it turn. Switching large currents on and off needs a special heavy-duty switch. If you had such a large switch inside the car it would be a nuisance since it would take up a lot of space. The switch would spark and it would be unpleasant and dangerous. A way round this problem is to use a **relay**.

TIP

Electromagnets are made with a soft iron core. Here soft is used to mean that the iron is a temporary magnet that is easily magnetised. Soft iron loses its magnetism quickly when the magnetising field is removed.

KEY TERM

A **relay** is a device which uses a small current to control a much larger current in a different circuit.

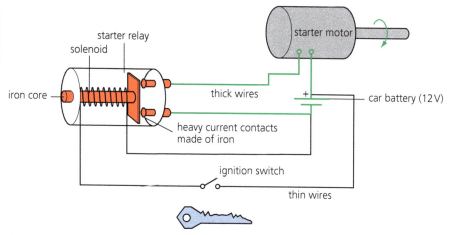

▲ **Figure 7.18** How a car starter relay works.

Inside the relay a solenoid is wound round an iron core. When the car ignition is turned, a small current magnetises the solenoid and its iron core. The solenoid is attracted towards the heavy-duty electrical contacts, which are also made of iron. Now current can flow from the battery to the starter motor. This system allows the car engine to be started by turning a key or pressing a switch at a safe distance.

Test yourself

5 List four ways to increase the strength of an electromagnet.

6 Why must you use insulated wire to make an electromagnet?

7 Sort these sentences into the correct order to explain how the relay operates to connect the car starter motor to the battery in Figure 7.18.

   A The ignition switch is turned.

   B The motor is connected to the battery.

   C The iron core becomes magnetised.

   D A current flows through the solenoid.

   E The solenoid moves to connect to the contacts.

▲ Figure 7.19

8 Figure 7.19 represents a wire placed vertically, with the current flowing out of the paper towards you. Copy the diagram and draw the magnetic field lines round the wire, showing how the field strength decreases with distance away from the wire.

9 Figure 7.20 shows a long Perspex tube with wire wrapped round it to make a solenoid.

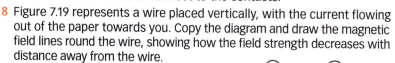

   a) Copy the diagram and mark in the direction of the compass needles 1–6 when the current flows through the wire.

▲ **Figure 7.20**

   b) Which end of the solenoid acts as the south pole?

   c) What happens when the current is reversed?

   d) Copy the diagram again, leaving out the compasses. Sketch the shape of the magnetic field.

Show you can...

Complete this task to show you understand about electromagnets.

Explain how you would use an iron nail, insulated wire and a cell to make an electromagnet that can be used to pick up some steel paper clips.

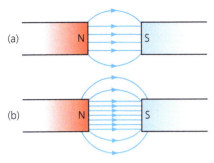

(a)

(b)

▲ **Figure 7.21** Lines of magnetic flux between two pairs of magnets. The flux density is greater in (b) than in (a).

KEY TERM ⭐

Flux density The number of lines of magnetic flux in a given area.

◯ Magnetic flux density

We represent magnetic fields by drawing lines that show the direction of a force on a north pole. These lines are also known as lines of **magnetic flux**.

Figure 7.21 shows the lines of magnetic flux between two pairs of magnets. The magnets in Figure 7.21b) are stronger than the magnets in Figure 7.21a). This means that they will exert a stronger attractive force on a magnetic material such as an iron nail. We show that the magnets are stronger by drawing more lines of magnetic flux for the area of the magnets.

The strength of the magnetic force is determined by the **flux density**, *B*.

The flux density is measured in tesla, T. A laboratory bar magnet produces a flux density of about 0.1 T near to its poles.

Calculating the force on a wire

The force on a wire of length *L* at right angles to a magnetic field and carrying a current, *I*, is given by the equation *F = BIL*:

$$F = BIL$$

$$\text{force} = \text{magnetic flux density} \times \text{current} \times \text{length}$$

where force is in newtons, N
magnetic flux density is in tesla, T
current is in amperes, A
length is in metres, m.

Example

In Figure 7.22 the wire carries a current of 3.0A. Calculate the force acting on the wire.

Answer

$$F = BIL$$
$$= 0.2 \times 3.0 \times 0.15$$
$$= 0.09\,\text{N}$$

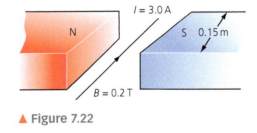

I = 3.0 A
N
S   0.15 m
B = 0.2 T

▲ **Figure 7.22**

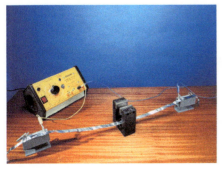

▲ **Figure 7.23** The aluminium foil carrying a current is pushed out of the magnetic field.

The motor effect

In Figure 7.23 you can see a piece of aluminium foil between the poles of a strong magnet. A current through the foil has caused it to be pushed down, away from the poles of the magnet. Reversing the current would make the foil move upwards, again away from the poles of the magnet. This is called the **motor effect**. It happens because of an interaction between the two magnetic fields: one from the permanent magnet and one produced by the current in the foil.

Combining two magnetic fields

In Figure 7.24 you can see the way in which the two fields combine. By itself the field between the poles of the magnet would be of constant strength and direction. The current around the foil produces a circular magnetic field. In one direction the magnetic field from the current squashes the field between the poles of the magnet. It is the squashing of the field that produces a force on the foil, upwards in this case.

The size of the force acting on the foil depends on:

- the magnetic flux density between the poles
- the size of the current
- the length of the foil between the poles.

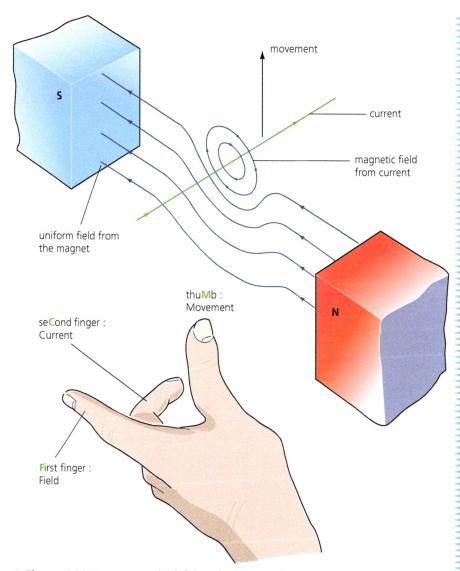

▲ **Figure 7.24** You can use the left-hand rule to predict the direction of movement of the wire.

The left-hand rule

To predict the direction in which a straight conductor moves in a magnetic field you can use **Fleming's left-hand rule** (Figure 7.24). Spread out the first two fingers and the thumb of your left hand so that they are at right angles to each other. Let your first finger point along the direction of the magnet's field (north to south), and your second finger point in the direction of the current (positive to negative). Your thumb then points in the direction in which the wire moves.

This rule works when the field and the current are at right angles to each other. When the field and the current are parallel to each other, there is no force on the wire and it stays where it is.

Try using the left-hand rule to show that the direction of the force on the conductor will reverse if:

- the direction of the magnetic field is reversed
- the direction of the current is reversed.

Test yourself

10 State two ways of increasing the force on a conductor carrying a current in a magnetic field.

11 How is it possible to position a wire carrying a current in a magnetic field so that there is no force acting on the wire?

12 Use the left-hand rule to predict the direction of the force on the wire in each of the following cases.

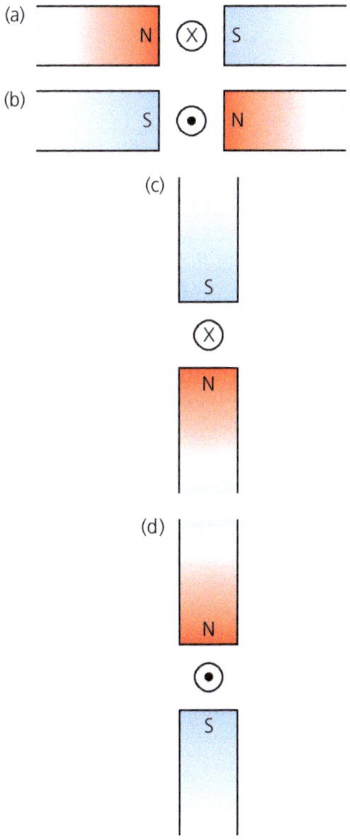

(a)

(b)

(c)

(d)

▲ Figure 7.25

13 A wire of length 0.2 m is placed at right angles to a region with a magnetic flux density 2.0 T. A current of 4.5 A is passed through the wire. Calculate the force acting on the wire.

TIP ✓

In an exam you will only be expected to use the left hand rule to help you recall that the magnetic field current and the movement of the wire are all at right angles to each other.

Electric motor

Figure 7.26 shows a coil of wire that is able to rotate about an axle. When a current flows, as shown in the diagram, there is an downward force on the side CD and a upward force on the side AB. So the coil begins to rotate clockwise, but the coil will rotate no further than a vertical position.

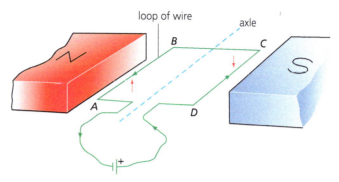

▲ **Figure 7.26** When the current flows, the coil begins to rotate clockwise.

Figure 7.27 shows how we can design a motor so that direct current can keep a coil rotating all the time.

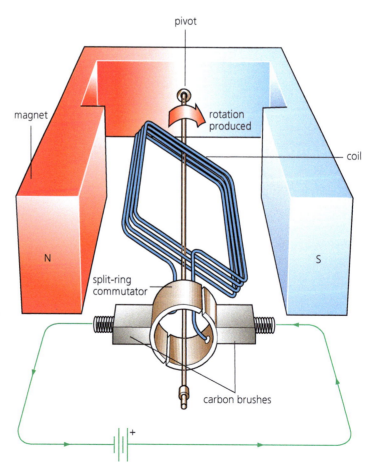

▲ **Figure 7.27** A simple motor.

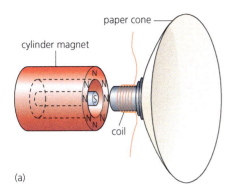

paper cone

cylinder magnet

coil

(a)

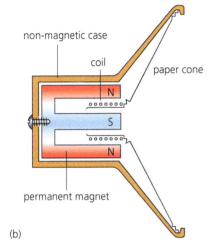

non-magnetic case

coil

paper cone

permanent magnet

(b)

▲ **Figure 7.28** Two views of a loudspeaker.

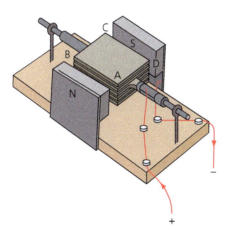

▲ **Figure 7.29** A model electric motor.

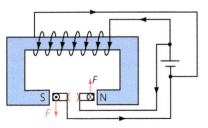

▲ **Figure 7.30**

The coil is kept rotating by using a **split-ring commutator**, which rotates with the coil between the carbon brush contacts. As the coil passes the vertical position, the two halves of the commutator change contact from one carbon brush to the other. This causes the direction of the current in the coil to reverse, so that the forces continue to act on the coil in the same direction. The coil will continue to rotate clockwise as long as there is a current.

The moving-coil loudspeaker

You can check the forces on the loudspeaker coil by looking at Figure 7.28(b). When the current in the upper side of the coil is flowing into the paper, the current in the lower side of the coil flows out of the paper. If you now apply the left hand rule to both sides of the coil, you will find that both sides experience a force to the left. When the current is reversed, the coil experiences a force to the right.

Test yourself

14 Look at Figure 7.26. Explain why there is no force acting on the side BC.

15 Give three ways in which you can increase the speed of rotation of the coil of an electric motor.

16 This question refers to the loudspeaker in Figure 7.28. In part (b) the current flows into the page as we look at the top of the coil and out of the page at the bottom.
   a) What happens to the coil when the direction of the current is reversed?
   b) Explain what happens to the coil when an a.c. current is supplied to it.
   c) What happens to the sound produced by the loudspeaker when these changes are made to an a.c. supply to the coil:
      i) the frequency is increased
      ii) the size of the current is increased?

17 Figure 7.29 shows a model electric motor. In the diagram the current flows round the coil in the direction A to B to C to D.
   a) What is the direction of the force on
      i) side AB
      ii) side CD?
   b) In which direction will the coil rotate?
   c) In which position is the coil most likely to stop?

18 Figure 7.30 shows an electric motor that is made using an electromagnet. The arrows show the direction of the forces on the coil.
   a) The battery is now reversed. What effect does this have on:
      i) the polarity of the magnet
      ii) the direction of the current in the coil
      iii) the forces acting on the coil?
   b) Explain why this motor works using an a.c. supply as well as a d.c. supply.
   c) Why are the electromagnet and coil run in parallel from the supply rather than in series?

Show you can...

Show you understand the purpose of the parts of a motor by explaining how the split-ring commutator keeps an electric motor spinning in the same direction all the time.

Induced potential, transformers and the National Grid

○ Induced potential

When a conducting wire moves through a magnetic field, a potential difference (p.d.) is produced across the ends of the wire. A potential difference produced in this way is described as an **induced potential difference**. A potential difference will also be induced if there is a change in the magnetic field around a stationary conducting wire.

If the wire is part of a complete circuit, the potential difference will cause a current in the circuit. This is an **induced current**.

Changing the direction of motion of the conducting wire or changing the polarity of the magnetic field will change the direction of the induced p.d. and reverse the direction of the induced current.

Inducing a p.d. or a current in this way is called the **generator effect**, because this is how we generate electricity.

Practical

Investigating induced potential differences

Figure 7.31 shows how you can investigate induced potential difference.

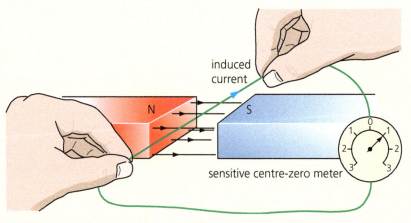

induced current

N   S

sensitive centre-zero meter

▲ Figure 7.31

To induce a p.d. across the ends of the wire, the wire must be moved up and down between the poles of the magnets. This is often described as the wire 'cutting' the magnetic field lines.

Method
Use the apparatus to show that the induced p.d. increases if:
- the wire is moved faster.
- stronger magnets are used.

Use the apparatus to check that when the the direction of the wire movement changes, from up to down, the direction of the induced p.d. changes.

○ Coils and magnets

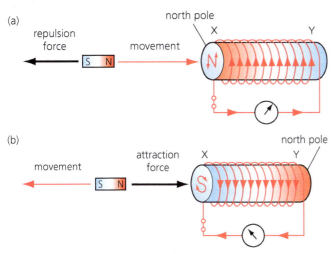

▲ **Figure 7.32** a) A north pole of a magnet is pushed into a solenoid. b) The magnet is pulled out of the solenoid.

We can also induce a p.d. in a coil of wire by changing the magnetic field near it. In Figure 7.32a) a north pole of a magnet is pushed into a solenoid. Note that the induced current flows in a direction that makes the end of the solenoid (X), next to the magnet, also behave like a north pole.

When the magnet is pulled out of the solenoid, the direction of the induced current is reversed. Now the end X behaves like a south pole.

When a current is induced in a conducting wire, the induced current itself generates a magnetic field. This magnetic field opposes whatever the original change was that caused it. This may be the movement of the conducting wire or the change in the magnetic field.

Consider Figures 7.32a) and b). In each case as the magnet is moved and a current induced, there is a force which opposes the movement of the magnet. So when you move a magnet to induce a current you do work.

An electromagnetic flow meter

Figure 7.33 shows a way to measure the rate of flow of oil through an oil pipeline. A small turbine is placed in the pipe, so that the oil flow turns the blades round. Some magnets have been placed in the rim of the turbine, so that they move past a solenoid. These moving magnets induce a p.d. in the solenoid. The p.d. can be measured on an oscilloscope (Figure 7.33b)). The faster the turbine rotates, the larger the p.d. induced in the solenoid. By measuring this p.d. an engineer can tell at what rate the oil is flowing.

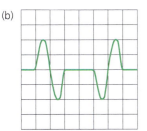

▲ **Figure 7.33** a) An electromagnetic flow meter. b) The induced p.d. changes as the magnets rotate.

> ### Test yourself
>
> 19 A wire is made to move through a magnetic field so that a potential difference is induced across its ends. State two ways in which the size of the induced potential difference can be increased.
> 20 This question refers to the magnet and solenoid shown in Figure 7.32.
>   a) The magnet is stationary inside the solenoid. State the size of the induced potential difference.
>   b) The magnet is now moved as follows. In each case state the direction of the deflection on the meter.
>   i) A south pole is moved towards X.
>   ii) A south pole is moved away from X.
>   iii) A north pole is moved towards Y.
> 21 This question refers to the flow meter shown in Figure 7.33.
>   a) The poles of the magnets on the wheel are arranged alternately with a north pole then south pole facing outwards. Use this fact to explain the shape of the trace on the oscilloscope.
>   b) Sketch the trace on the oscilloscope for these two separate changes:
>   i) the number of turns in the solenoid is doubled
>   ii) the oil flow rate slows down, so the turbine rotates at half the speed.

○ Uses of the generator effect

Figure 7.34a) shows the design of a simple electricity generator. This one generates alternating current, so it is called an **alternator**.

By turning the axle you can make a coil rotate in a magnetic field. This causes a p.d. to be induced between the ends of the coil.

Figure 7.34b) shows how the induced p.d. varies with time. This waveform can be displayed on an oscilloscope. Figure 7.34c) shows the position of the coil at the times marked on the oscilloscope.

i)   The coil is vertical. In this position the sides AB and CD are moving parallel to the field and no p.d. is induced.

ii) The coil is horizontal. In this position the sides AB and CD are cutting through the field at the greatest rate. The induced p.d. is at its maximum.

iii) The coil is again vertical and the induced p.d. is zero.

iv) The coil is again horizontal, but the sides AB and CD are moving in the opposite direction in comparison with position (ii). The induced p.d. is again at its maximum, but now in the opposite direction.

KEY TERM

The **slip-rings** provide a continuous contact between the rotating sides of the coil and the connections to the oscilloscope marked ① and ② in Figure 7.34

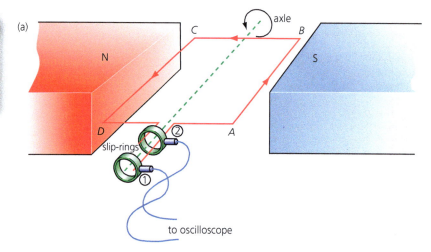

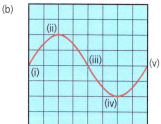

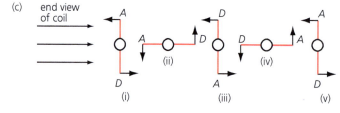

▲ Figure 7.34

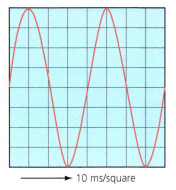

▲ **Figure 7.35** The waveform produced when the generator in Figure 7.34 is rotated twice as quickly.

The size of the induced p.d. can be made larger by:

- rotating the coil faster
- using stronger magnets
- using more turns of wire
- wrapping the wire round a soft iron core.

Figure 7.35 shows the potential difference waveform produced when the generator is rotated twice as quickly. There are two effects: the maximum potential difference is twice as large and the frequency is doubled, i.e. the interval between the peaks is halved.

○ The dynamo

Figure 7.36a) shows the design of a direct current generator. This is called a **dynamo**. The design of a dynamo is similar to that of an alternator. However, a dynamo has a split ring commutator rather than two separate slip rings. Now it does not matter which side, AB or CD, moves upwards, the induced current always flows in the same direction.

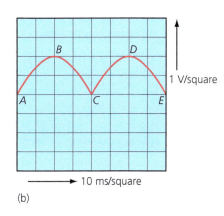

(b)

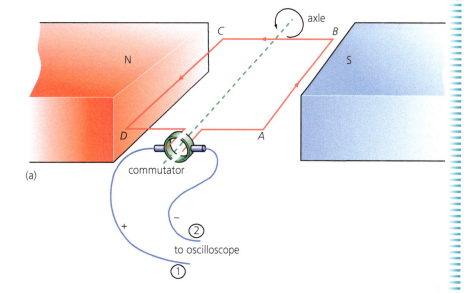

(a)

▲ **Figure 7.36** The design of a dynamo.

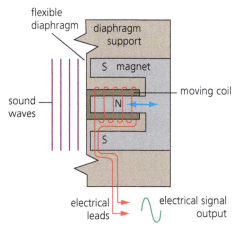

▲ **Figure 7.37** A moving coil microphone.

○ Microphones

Figure 7.37 shows how a moving coil microphone works. When sound waves reach the diaphragm they cause it to vibrate. The diaphragm is attached to a small moving coil, which then vibrates inside a strong magnetic field. In this way a p.d. is induced across the end of the coil.

When we use a microphone with a sound system, the induced p.d. is amplified and a current can then drive a loudspeaker, which enables an audience to hear the sound.

Show you can...

Show you understand how an alternator works by completing this task.

Explain how a coil and magnets can be used to generate alternating current. How can you make the induced current as large as possible?

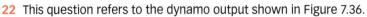

Test yourself

22 This question refers to the dynamo output shown in Figure 7.36.
   a) State the position of the coil for each of the times A, B, C, D and E.
   b) Copy the graph and add further graphs to show what happens to the p.d. after each of these separate changes.
      i)   The direction of the coil rotation is reversed.
      ii) An extra turn of wire is added.
      iii) The coil is rotated at twice the speed.
23 This question refers to the microphone shown in Figure 7.37. Explain why a larger p.d. is induced when the sound is louder.

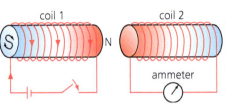

▲ Figure 7.38

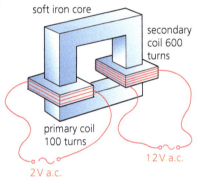

▲ Figure 7.39 A step-up transformer.

▲ Figure 7.40 Without transformers we could not transmit electricity over large distances.

◯ Transformers

Changing currents and fields

Figure 7.38 shows two coils placed close to each other. Coil 1 can be connected to a battery; coil 2 is connected to a sensitive ammeter.

- When the current is switched on, the ammeter in the second circuit moves to the right.
- When the current in coil 1 flows steadily, the ammeter reads zero.
- When the current is turned off in coil 1, the ammeter moves to the left.

Why does this happen?

This is rather like moving a magnet into a solenoid, then leaving the magnet stationary in the solenoid, then moving the magnet out of the solenoid again. A p.d. is induced one way when the magnet moves in and the other way when it moves out.

You can increase the size of the induced current further by putting an iron rod between the coils. Now the current in the first coil produces a larger changing magnetic field, which induces a larger current in the second coil.

A **transformer** is made by putting two coils of wire onto an iron core, as shown in Figure 7.39.

Transformers only work using an a.c. supply. When an alternating current is supplied to the primary coil, a changing magnetic field is produced. The iron core becomes magnetised and carries the changing magnetic field to the secondary coil. The changing magnetic field passes through the secondary coil where an alternating p.d. is induced.

When a direct current is supplied to the primary coil, the magnetic field is constant. Then no p.d. is induced in the secondary coil.

Step-up and step-down transformers

Transformers are useful because they allow us to change the p.d. of an a.c. supply. The transformer in Figure 7.39 is an example of a step-up transformer. The input p.d. is increased from 2 V to 12 V.

- When there are more turns on the secondary coil than the primary coil, the p.d. is increased. This is a **step-up** transformer.
- When there are fewer turns on the secondary coil than the primary coil, the p.d. is decreased. This is a **step-down** transformer.

Example

An a.c. power supply has a p.d. of 230V. This is applied to a transformer with 5000 turns in its primary coil and 250 turns in its secondary coil. Calculate the potential difference across the secondary coil. Is this a step-up or step-down transformer?

Answer

$$\frac{V_p}{V_s} = \frac{n_p}{n_s}$$

$$\frac{230}{V_s} = \frac{5000}{250}$$

$$\frac{230}{V_s} = 20$$

So $V_s = \dfrac{230\,V}{20} = 11.5\,V$

This is a step-down transformer.

The rule for calculating the potential differences in a transformer is:

$$\frac{V_p}{V_s} = \frac{n_p}{n_s}$$

where V_p is the p.d. across the primary coil

V_s is the p.d. across the secondary coil

n_p is the number of turns in the primary coil

n_s is the number of turns in the secondary coil.

Power in transformers

We use transformers to transfer electrical power from the primary coil to the secondary coil. Most transformers do this very efficiently and there is little loss of power in the transformer itself. For a transformer that is 100% efficient we can write:

power supplied to the = power provided by the
primary coil   secondary coil

$$V_p \times I_p = V_s \times I_s$$

where I_p is the current in the primary coil

I_s is the current in the secondary coil

V_p is the p.d. across the primary coil

V_s is the p.d. across the secondary coil

Practical

Investigating your own transformer

You can make and investigate your own transformer by wrapping two coils of insulated wire around iron C-cores. Connect one coil, the primary, to a 2V alternating power supply. Connect the second coil, the secondary, to a 3V lamp.

Without changing the number of turns on the primary coil or the potential difference across the primary coil, wrap more turns onto the secondary coil.

1 What happens to the brightness of the lamp as more turns are wrapped onto the secondary coil?

2 Why must the number of turns on the primary coil and the potential difference across the primary coil be kept constant?

3 What is the independent variable in this investigation?

4 Why should the number of turns on the secondary coil not go too far above the number of turns on the primary coil?

(Only switch the power supply on for a few seconds at a time.)

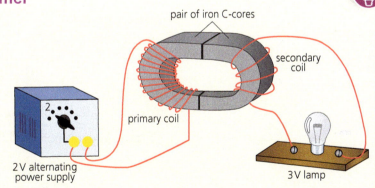

▲ Figure 7.41

○ **The National Grid**

Figure 7.42 shows how electricity generated in a power station is distributed around the country through the National Grid.

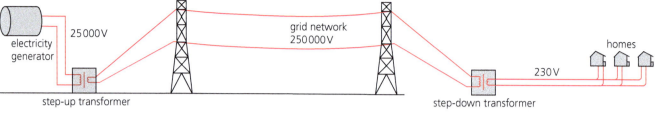

▲ **Figure 7.42** How power is supplied by the National Grid.

Power is transmitted around the country at a potential difference as high as 400 000 V. There is a very good reason for this. It makes the whole process much more efficient.

The following calculations explain why.

Figure 7.43 suggests two ways of transmitting 25 MW of power from a Yorkshire power station to the Midlands.

a) The 25 000 V supply from the power station could be used to send 1000 A down the power cables.

b) The potential difference could be stepped up to 250 000 V and 100 A could be sent along the cables.

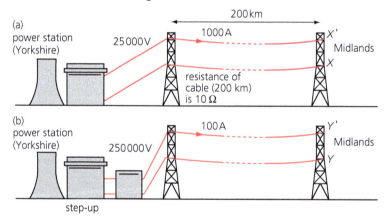

► **Figure 7.43** Two ways of transmitting 25 MW of power from a Yorkshire power station to the Midlands.

How much power would be wasted in heating the cables in each case given that 200 km of cable has a resistance of 10 Ω?

a) power lost = potential difference along cable × current

$$= IR \times I$$

$$= I^2 R$$

$$= (1000)^2 \times 10$$

$$= 10\,000\,000\,\text{W or } 10\,\text{MW}$$

b) power lost $= I^2 R$

$$= (100)^2 \times 10$$

$$= 100\,000\,\text{W or } 0.1\,\text{MW}$$

By stepping up the potential difference to a very high level, power is transmitted using much lower currents. When a high current flows through a wire, energy is used to heat the wire up. By keeping the current low, less energy is wasted and the whole system is much more efficient.

TIP

Electricity is transmitted at a very high potential difference and low currents to reduce energy loss and improve efficiency.

Test yourself

24 In Figure 7.38, when the switch is opened the ammeter flicks to the left, describe what happens to the ammeter during each of the following:

   a) The switch is closed and left closed so that a current flows through coil 1.

   b) The coils are pushed towards each other.

   c) The coils are left close together.

   d) The coils are pulled apart.

25 The table below gives some information about four transformers. Copy the table and fill in the gaps.

Primary turns	Secondary turns	Primary p.d. in volts	Secondary p.d. in volts	Step-up or step-down
100	20		3	
400	10 000	10		
	50	240	12	
	5 000	33 000	11 000	

26 Give three examples of where transformers are used at home to step the mains potential difference of 230 V down to a lower level.

27 Explain why a transformer does not work using a d.c. supply. Why is an a.c. supply necessary to make a transformer work?

28 Explain why very high potential differences are used to transmit electrical power over long distances.

Show you can...

Show you understand about the transmission of electricity around the country by explaining why we use a.c. electricity in our homes, rather than d.c. electricity supplies.

Chapter review questions

1 **a)** Draw accurately the magnetic field pattern around a bar magnet.

b) Explain how you would use a compass to plot this magnetic field pattern.

2 Draw carefully the shape of a magnetic field close to:

a) a long wire   **b)** a long solenoid

when each is carrying a current.

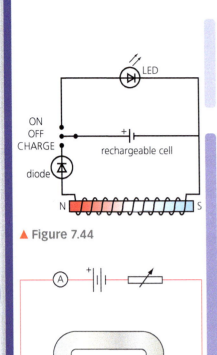

▲ Figure 7.44

3 Figure 7.44 is a circuit diagram for a 'shake-up' torch. A magnet is placed in a coil of wire, which is linked to a rechargeable battery and a light-emitting diode (LED). The magnet is free to slide backwards and forwards inside the coil.

a) What is a light-emitting diode?

b) Explain why a diode is used in the charging circuit.

c) Explain how the cell is recharged.

d) Explain why it is necessary to have the 3 switch positions marked on, off, charge.

4 Figure 7.45 shows an experimental arrangement to investigate the action of an electromagnet. Coils of wire have been wound round a C-shaped soft iron core.

At the bottom of the C-core, a soft iron bar stays in place because of the attraction of the electromagnet.

The strength of the electromagnet is measured by increasing the weight hanging on the bar until it falls off.

The table below shows some results for this experiment.

Maximum weight on the iron bar in N	0	0.6	1.2	1.8	2.3	2.7	3.0	3.2	3.3	3.4	3.4
Current in the magnet coils in A	0	0.2	0.4	0.6	0.8	1.0	1.2	1.4	1.6	1.8	2.0

▲ Figure 7.45

a) Plot a graph of the maximum weight supported by the electromagnet (*y*-axis) against the current in the magnet coils (*x*-axis).

A second student does the same experiment, but uses more turns of wire on her electromagnet.

b) Sketch a second graph, using the same axes, to show how the maximum weight supported changes with the current now.

5 Figure 7.46a) shows a bicycle dynamo. When a bicycle wheel moves it turns the driving wheel of the dynamo. The driving wheel is attached to a permanent magnet which rotates inside a fixed coil.

The output terminals of the dynamo are connected to a data logger, which displays the potential difference shown in Figure 7.46b).

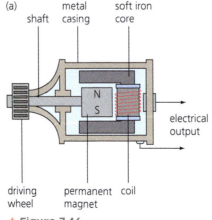
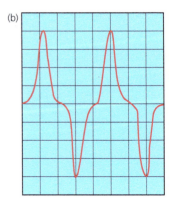

▲ Figure 7.46

When the potential difference looks as it does in Figure 7.47(b), the drive wheel is rotating 10 times each second.

a) Copy Figure 7.46b) and add to it a second trace to show the potential difference when the wheel rotates 5 times each second.

b) Explain why a bicycle lamp attached to the dynamo does not light when the bicycle is stationary.

6 Figure 7.47 shows a laboratory demonstration of how a transformer may be used to produce very large currents. Here the current is used to melt a nail.

a) Use the information in the diagram to show that the potential difference across the nail is 2.3 V.

b) The nail has a resistance of 0.02 Ω. Calculate the current in the secondary circuit.

c) Show that the power in the secondary circuit is 265 W.

The nail melts when 15 000 J of energy have been transferred to its thermal store.

d) Calculate how long it takes for the nail to melt.

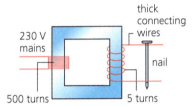

▲ **Figure 7.47**

7 The generator in a power station produces electricity at a potential difference of 25 000 V. A transformer steps this up to 400 000 V for the electricity to be transmitted on the National Grid.

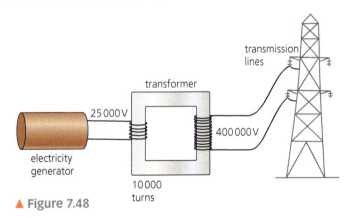

▲ **Figure 7.48**

a) Use the information in the diagram to calculate the number of turns in the secondary coil of the transformer.

b) The current in the primary coil is 800 A. Calculate the current in the transmission lines.

c) Why are high potential differences used to transmit current on the National Grid?

Practice questions

1 The diagram shows a magnetic screwdriver that has picked up and is holding a small metal screw.

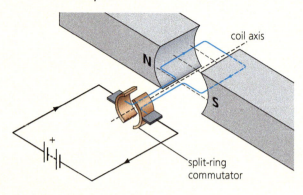

a) What type of material is the screw made from? [1 mark]

aluminium   copper   steel

b) The magnetic force between the screw driver and screw is an example of a non-contact force.

Which **one** of the following is also a non-contact force?

air resistance   friction   gravity   [1 mark]

▲ Figure 7.49

c) Explain how you can use a bar magnet to show that an unmarked bar of material is either a bar magnet or an unmagnetised bar of iron. [3 marks]

2 Figure 7.50 represents a simple transformer used to light a 12 V lamp. When the a.c. input is supplied to the primary coil, the lamp is dim.

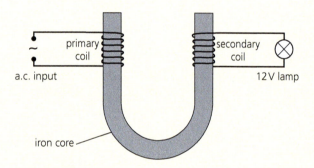

▲ Figure 7.50

a) Copy and complete the sentences below. [5 marks]

The alternating current input in the _____ coil produces a continuously changing _____ _____ in the iron core and therefore through the secondary coil. This _____ an alternating potential difference across the ends of the secondary coil. If the secondary coil is part of a complete circuit, _____ will be induced in the _____ coil.

b) Suggest three ways to increase the potential difference across the lamp without changing the power supply. [3 marks]

3 Figure 7.51 shows a diagram of an electric motor. As the current flows from the battery, the coil and split-ring commutator spin.

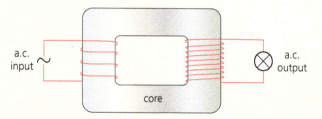

▲ Figure 7.51

a) Explain why the coil spins when a current flows through it. [3 marks]

b) Without changing the coil, give two ways in which it could be made to spin faster. [2 marks]

c) Give two ways in which the coil could be made to spin in the opposite direction. [2 marks]

4 Figure 7.52 shows a simple transformer.

a.c. input ~ ... core ... a.c. output

▲ Figure 7.52

a) Is the transformer being used as a step-up or step-down transformer? [1 mark]

b) Why must the wire used to make the coils be insulated? [1 mark]

c) i) What material is the core made from?

ii) Why must the core be made from this material? [2 marks]

5 Figure 7.53 shows some apparatus that is being used to investigate the factors which affect the potential difference induced when the wire CD moves through a magnetic field. The potential difference is measured by a sensitive voltmeter which reads zero when the pointer is in the middle of the scale.

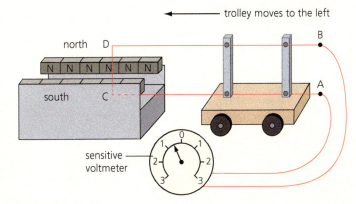

▲ **Figure 7.53**

The trolley is moving to the left and the sensitive voltmeter records a reading to the left of zero.

a) State **two** separate changes you could make so that the voltmeter records a reading to the right of zero. [2 marks]

b) State **two** separate changes you could make so that the induced potential difference is greater. [2 marks]

c) What does the voltmeter record when the wire CD is stationary between the magnets? [1 mark]

6 a) A student investigates how a thick copper wire can be made to move in a magnetic field. Figure 7.54 shows the apparatus.

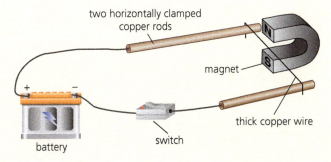

▲ **Figure 7.54**

The wire is placed between the poles of the magnet.

i) Use the information in the diagram to predict the direction of motion of the wire. [1 mark]

ii) Explain what happens to the motion of the wire when the magnet is turned so the N pole is below the wire. [2 marks]

b) Figure 7.55 shows a model generator.

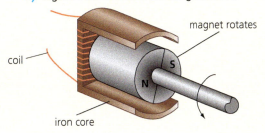

▲ **Figure 7.55**

i) Explain why a potential difference is induced across the ends of the coil when the magnet rotates. [2 marks]

ii) Explain why the potential difference is alternating. [1 mark]

c) The ends of the coil are connected to a cathode ray oscilloscope (CRO). Figure 7.56 shows the trace on the screen as the magnet rotates.

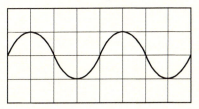

▲ **Figure 7.56**

Copy the diagram and draw new traces for each of the following changes using the same scale. The settings of the oscilloscope remain the same.

i) The magnet rotates at the same speed, but in the opposite direction. [1 mark]

ii) The magnet rotates at the same speed, in the same direction as the original, but the number of turns of the coil is doubled. [2 marks]

iii) The magnet rotates at twice the speed, in the same direction, with the original number of turns of the coil. [2 marks]

d) Explain why iron is used as the core in the model generator. [1 mark]

7 Figure 7.57 shows part of a bicycle dynamo which is in contact with the wheel.

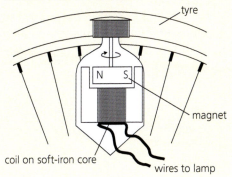

▲ **Figure 7.57**

a) Explain fully why a current flows through the lamp when the bicycle wheel turns. [3 marks]

b) Why does the lamp get brighter as the cycle moves faster? [2 marks]

c) Why does the lamp not work when the bicycle is stationary? [1 mark]

8 Figure 7.58 shows a type of door lock.

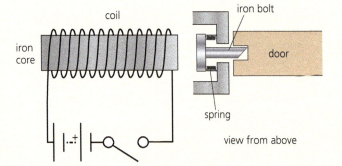

▲ Figure 7.58

a) Explain how closing the switch allows the door to be opened. [3 marks]

b) The door bolt is changed, and a thicker, stronger piece of iron is used. When the switch is closed, the lock does not open. Without changing the bolt, suggest two changes that could be made, each of which would make the lock work again. [2 marks]

9 Figure 7.59(a) shows a coil connected to a sensitive meter. The meter is a 'centre zero' type: the needle is in the centre when no current flows.

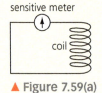

▲ Figure 7.59(a)

A student does two experiments.

In the first experiment, shown in Figure 7.59(b), the magnet is pushed into the coil. Then the magnet is removed. The meter only reads a current when the magnet is moving.

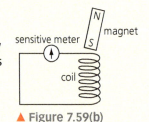

▲ Figure 7.59(b)

In the second experiment, shown in Figure 7.59(c), a second coil is brought close to the first. The student switches the current on and off in the second coil. The needle deflects for a brief time each time the current is turned on or off.

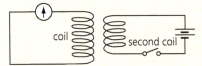

▲ Figure 7.59(c)

a) In the first experiment:

i) What is happening in the coil while the magnet is moving? [1 mark]

ii) How does the deflection of the needle as the magnet is pushed towards the coil compare with the deflection of the needle as the magnet is pulled away? [1 mark]

iii) State three ways in which the student could increase the deflection on the meter when the magnet is moved. [3 marks]

b) Use the second experiment to help explain why transformers only work with alternating current. [3 marks]

10 The waves from earthquakes are detected by instruments called seismometers. Figure 7.60 shows a simple seismometer.

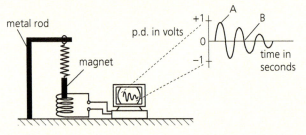

▲ Figure 7.60

It consists of a bar magnet suspended on a spring. The spring hangs from a metal rod that transmits vibrations from the Earth. When there is an earthquake, the magnet moves in and out of the coil. A computer monitors the potential difference across the coil.

a) Explain why a p.d. is induced in the coil. [1 mark]

b) Why is the induced p.d. alternating? [1 mark]

c) Describe the movement of the magnet when the induced p.d. has its greatest value at the point labelled A. [1 mark]

d) Describe the movement of the magnet when the induced p.d. is zero, as at point B. [1 mark]

e) Suggest **two** ways in which the seismometer could be made more sensitive so that it can detect weaker earthquakes. [2 marks]

Working scientifically

Technological applications of science: the implications

The discovery of the principles of electromagnetic induction by Michael Faraday led to the development of the d.c. dynamo. The dynamo was the first machine capable of generating electricity for industrial processes.

Further scientific work led to the development of the a.c. generator. This, combined with the invention of the transformer, resulted in dynamos being replaced by a.c. generators for electricity generation and power distribution.

Easy access to electricity has led to more appliances, devices and gadgets being invented that use electricity. However, easy access to electricity is not a reality for everyone. For about 1.6 billion people in the world, life without electricity is a reality. There is no flicking a switch for light, pressing a button to work the microwave, or putting dirty clothes in a washing machine.

Having electricity is not just about being rich or poor. It is about a country having the resources to generate electricity or the money to pay for electricity. The use of the world's resources to provide some people with a comfortable lifestyle whilst others go without raises **ethical issues**.

The application of science to everyday life and to technology may also have certain personal, social, economic and environmental implications. Consider the examples of the clockwork radio and water bottle lighting system.

The clockwork radio

The clockwork radio was invented by Trevor Baylis in 1989. The radio was developed for use in countries where affordable energy is either rare or does not exist. The original radios had no batteries and did not need to be plugged into the mains electricity supply. Turning the handle powered a small internal clockwork generator. This provided the electricity to work the radio.

Most people turning a generator like the one in the wind-up radio are able to develop about 5 watts of power. This is an energy input to the generator of 5 joules every second.

The table shows the energy required to operate a few devices for just 1 minute.

Device	Energy to operate for 1 minute in joules
MP3 player	60
Mobile phone	120
Desktop computer	12 000
Electric kettle	180 000

Questions

1 Why was the invention of the transformer important in developing the electrical power distribution systems used today?
2 Why does the increased use of electricity in developed countries raise ethical issues?

KEY TERM

An **ethical issue** involves a consideration of what is good or bad for both individuals and society as a whole. For many ethical issues there may be no single right or wrong answer.

▲ Figure 7.61 A clockwork radio.

Questions

3 Suggest what the personal and social implications of developing a clockwork radio may be.

4 Explain why it is unlikely that a wind-up desktop computer or electric kettle would be developed.

The water bottle lighting system

Using a plastic bottle filled with water (and a little bleach) to light up a dark room is a new idea that uses simple physics.

▲ **Figure 7.62** A plastic bottle filled with water provides about the same amount of light as a 50W bulb.

The water bottle fits into a hole made in the roof of the building. Sunlight enters the bottle from above, is refracted and then spreads into the room. It is a simple device that, for a small cost, lets people with no electricity, or little money to pay for electricity, see inside a room that has no other source of daylight.

Question

5 Suggest one economic and one environmental implication of using this water bottle lighting system.

8 Space physics

People have had a fascination for stars and planets since the dawn of civilisation thousands of years ago. Our interest is so great that we have landed men on the Moon and sent space probes to all the planets in the Solar System – all at great cost. In 2015 NASA's New Horizon space probe reached the dwarf planet Pluto and its five moons.

Specification coverage

This chapter covers the specification point: 4.8 Space physics. It covers, 4.8.1 Solar system, stability of orbital motions, satellites and 4.8.2 Red shift.

Prior knowledge

Previously you could have learned

> The Sun is a star that radiates energy and light. There are many other stars in the sky.
> The Earth is a planet which orbits the Sun.
> The Moon is a natural satellite that orbits around the Earth, approximately once a month.
> There are eight planets and several dwarf planets which orbit around the Sun.

▲ **Figure 8.1** a) Saturn with its rings; b) Earth from space, showing Antarctica.

Test yourself on prior knowledge

1 What force keeps Saturn's rings in position, as shown in Figure 8.1a)?
2 Explain the difference between a star and a planet.

The Solar System

○ Planets

KEY TERM
A **planet** is a large body which orbits the Sun.

The brightest object you can see in the sky is our Sun. The Sun is a star which releases energy by the process of nuclear fusion. The Earth is a **planet** which moves around the Sun in a path called an orbit. In total there are eight planets, which are kept moving in their orbits by the pull of the Sun's gravitational force.

Unlike the Sun, planets do not produce their own light. We see planets because they reflect the Sun's light. The Sun, planets, dwarf planets and moons are known as the Solar System.

The inner planets

The four inner planets of our Solar System (in order out from the Sun) are Mercury, Venus, Earth and Mars. All these planets have hard, solid, rocky surfaces.

KEY TERM
A **moon** is a natural satellite in orbit around a planet.

▲ **Figure 8.2** Jupiter and four of its moons.

The outer planets

Beyond the orbit of Mars are four giant planets: Jupiter, Saturn, Uranus and Neptune. All these planets are much larger than the inner planets. The outer planets are made of gas and a spacecraft would sink if it tried to land on one.

A **moon** is a natural satellite that orbits a planet. The Earth is not the only planet with a moon. Jupiter, the largest planet, has 66 known moons. The photograph in Figure 8.2 shows four of Jupiter's moons which orbit round the planet.

Ganymede, Jupiter's largest moon, has a diameter of 5300 km. This is larger than any dwarf planet.

The other giant planets also have large numbers of moons. It seems likely that their strong gravitational pull attracted a lot of rocky material billions of years ago, as the Solar System formed. The table gives information about the planets in our Solar System.

Information about planets in our Solar System.

Planet	Average distance of planet from the Sun	Time taken to orbit the Sun	Number of moons
Mercury	58 million km	88 days	0
Venus	108 million km	225 days	0
Earth	150 million km	365 ¼ days	1
Mars	228 million km	687 days	2
Jupiter	780 million km	12 years	66
Saturn	1430 million km	29 years	62
Uranus	2880 million km	84 years	27
Neptune	4500 million km	165 years	13

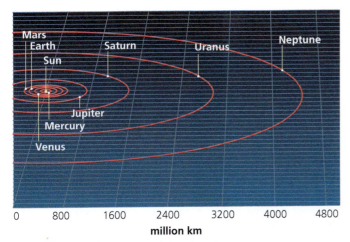

▲ **Figure 8.3** The orbits of planets around the Sun.

Dwarf planets

In 1930, Pluto was discovered and it was named as our ninth planet because it orbits the Sun. However, by 2005 several other objects had been discovered beyond Pluto's orbit, and one of them, Eris, is more massive than Pluto. These discoveries led to Pluto (and similar objects) being classified as **dwarf planets**.

A dwarf planet is large enough for its own gravity to have shaped it as roughly spherical. There are five objects now accepted as dwarf planets: Ceres, Pluto, Haumea, Makemake and Eris. It is highly likely that more dwarf planets will be discovered in the future.

The pull of gravity and orbits

You should understand that gravity acts close to the Earth to pull things to the ground. However, gravity is also a force that stretches an infinite distance out into space. Gravity acts on all stars, planets and moons.

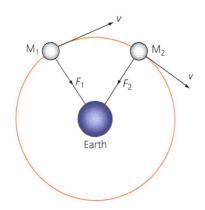

▲ **Figure 8.4** The force of gravity directed towards the Earth keeps the Moon in orbit.

- Gravity causes planets to orbit the Sun.
- Gravity causes the Moon and artificial satellites to orbit the Earth.
- Gravity causes stars to orbit around the centre of their galaxies.

Figure 8.4 shows the pull of gravity from the Earth, which keeps the Moon in orbit. In position M_1 the Moon has a speed v. The pull of gravity deflects its motion. Later it has moved to its new position M_2, but it still has the same speed, v. The force of gravity does not make the Moon travel any faster, but the force changes the direction of motion. The Moon will stay in its orbit for billions of years.

○ Galaxies

When you look at the sky at night you can see thousands of stars. Our Sun belongs to a large group of stars called a **galaxy**. We call our galaxy the Milky Way; it contains over 200 000 million stars.

▲ **Figure 8.5** Our galaxy, the Milky Way.

If you could see our galaxy from the side, it would look like two fried eggs stuck back to back (Figure 8.6a); it is long and thin except for a bulge in the middle. If you could see the galaxy from the top it would look like a giant whirlpool with great spiral arms (Figure 8.6b). In fact, the galaxy does spin round. Our Sun takes about 220 million years to go once round the centre of the galaxy. In your lifetime, the pattern of stars that you see each night will not appear to change. But over thousands of years the pattern will change as our Sun moves through the galaxy.

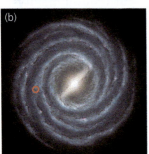

▲ **Figure 8.6** An artist's impression of our galaxy, the Milky Way. The red circle shows the approximate position of our Sun.

Practical

a) Use the table on page 249 to help you construct a model of the Solar System. The scale for your model is to be 1 m for 20 million kilometres. Using the data in the table calculate how far you must place each planet away from the Sun in your model. Now go out and persuade someone to represent each planet – you will need a lot of space.

b) Now work out where to place the next nearest star on your model. This is Proxima Centauri. Should the star be placed in the next street, the next town or where? (Proxima Centauri is about 9000 times further from the Sun than Neptune.)

Test yourself

1 Use words from the box to complete each sentence. Copy the completed sentences out.

> planet   dwarf planet   moon   star   galaxy

  a) A _____ orbits the Sun and a _____ orbits a planet.
  b) A _____ is a large group of stars.
  c) Pluto is now classified as a _____.

2 Explain the difference between a moon and a dwarf planet.

3 What are the differences between the four planets closest to the Sun and the four planets furthest away from the Sun?

4 a) What is the Milky Way?
  b) Make a sketch of our galaxy.
  c) Find out the names of two other galaxies.

5 Pluto used to be called a planet. In 2006, it was renamed as a dwarf planet. Explain why.

Show you can...

Complete this task to show you understand the structure of our Solar System.

Name four types of body that exist in our Solar System. Draw a diagram to explain their positions and how they move relative to each other.

○ Life cycle of a star

Our Sun is a star which was formed about 4.6 billion years ago; the Earth and planets were formed at about the same time. Figure 8.7 illustrates how we think stars are formed.

Birth of a star

Figure 8.7a) shows the Eagle Nebula. This is a cloud of cold hydrogen gas and dust which is collapsing due to the pull of gravity. As the cloud collapses, the atoms and molecules move very fast. As molecules collide with each other, their kinetic energy is transferred to the internal energy of the gas, and the temperature rises to several million degrees Centigrade. The contracting and heating ball of gas is called a **protostar**.

KEY TERM

A **protostar** is the name given to the large ball of gas as it contracts to form a star.

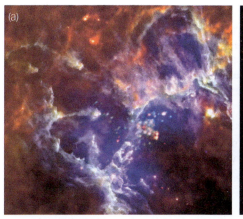

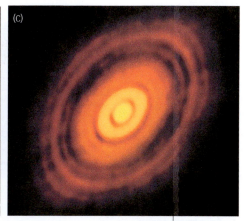

▲ **Figure 8.7** a) Stars begin to form from giant clouds of dust and hydrogen. b) A young star has been born. c) A solar system condenses. This is the new star H L Tauri, 450 light years away from our Solar System.

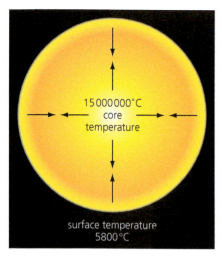

▲ **Figure 8.8** The outward force due to the fusion energy trying to expand a star is balanced by the inward gravitational force trying to collapse a star.

KEY TERMS ★

A **main sequence** star is one that releases energy by fusing hydrogen to form helium.

A **red giant** is a very large star which fuses helium into heavier elements.

At the end of its life, after fusion stops, a main sequence star collapses into a **white dwarf**.

The temperature in the gas becomes so high that hydrogen nuclei (protons) begin to collide, and fusion begins. Now a star has been born which releases energy for millions of years from nuclear fusion. Figure 8.7b) shows a star which has just begun to shine out from its cloud of dust and hydrogen.

The stable period of a star

The Sun is about half way through its life of about 10 billion years. For most of its life the Sun will release a constant output of energy. The Sun is described as a **main sequence** star, which means that it releases energy from the fusion of hydrogen to form helium. The inside of a star is a battleground (Figure 8.8). The forces of gravity tend to collapse the star. The enormous pressure of the fusion energy trying to expand the star balances these inward forces. The pressure at the centre of the Sun is about 100 billion times greater than atmospheric pressure.

Main sequence stars vary considerably in their masses and their brightness. The brightest main sequence stars are a million times brighter than the Sun, and the dullest about 1000 times duller.

The death of a star

Stars live for a very long time, but eventually their life cycle comes to an end. The final stages in the life cycle are determined by the size of the star.

Stars about the size of the Sun

Towards the end of a star's life as a main sequence star, the supply of hydrogen begins to run out. Now the star becomes unstable. Without the fusion of hydrogen, the pressure inside the star drops, the outward forces decrease, and the star begins to collapse. As the star collapses, the temperature of the core increases even further, reaching as high as 100 million °C. At such high temperatures helium begins to fuse to make heavier elements such as carbon and oxygen. The hot core causes the star to swell up into a **red giant**, which has about 100 times the radius of our Sun and about 1000 times the brightness. It is possible that the Sun might grow so large that it swallows up the Earth.

Eventually the star is no longer able to fuse helium. At this stage the star collapses into a **white dwarf** star, not much larger than the Earth. The surface of a white dwarf is hot with a temperature of 50 000 °C or more. Fusion stops and the star's life is over. At this stage the white dwarf star cools down and it will eventually become a dark cold star known as a **black dwarf**.

Stars much larger than the Sun

Large stars – with a mass of over 10 times that of the Sun – have much more dramatic ends to their lives. When such a star reaches the end of its main sequence stage, it too begins to collapse.

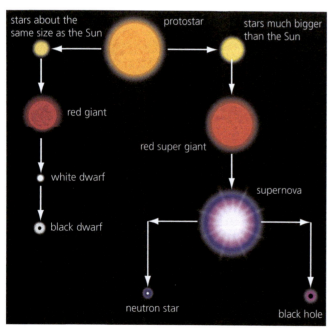

▲ **Figure 8.9**

But now the star grows into a **red super giant**, which could have a radius 1000 times that of the Sun and be 100 000 times brighter than the Sun. The fusion energy in red super giants is sufficient for heavier elements to be made in the fusion process. Iron is the heaviest element made inside stars due to fusion.

After millions of years as a red super giant, the star runs out of its nuclear fuel. There is a very rapid collapse of the star and two things happen.

- The rapid collapse heats the inside of the star to such high temperatures that there is a runaway nuclear reaction. The star explodes like a cosmic nuclear bomb. Such is the energy that elements heavier than iron are made. The remnants of a supernova are spread out into space. Eventually these remnants form part of another cloud of gas, which collapses to form new stars. Our planet has elements heavier than iron – so our world was once part of another star.
- At the same time the great gravitational forces cause the centre to collapse into a highly condensed form of matter. The core might be left as a neutron star a few kilometres across, which is made only out of neutrons. Or the collapse can be so complete that the star disappears into a microscopic point and it has become a black hole. A black hole is so dense that nothing, including light, can escape from it.

▲ **Figure 8.10** The Crab Nebula is the remnant of a supernova which was recorded by Chinese astronomers in 1054.

KEY TERMS

A **supernova** is a gigantic explosion caused by runaway fusion reactions in a very large star.

A **neutron star** is a very dense small star made out of neutrons.

A **black hole** is the most concentrated state of matter, from which even light cannot escape.

Show you can...

Show you understand the life cycle of a star by completing this task.

Write an account of the life cycle of our Sun from its birth to its end.

Test yourself

6 **a)** The various stages of the life cycle of a star like the Sun are listed below. Put them in the correct order.
   - **A** Main sequence star
   - **B** White dwarf
   - **C** Red giant
   - **D** Protostar

   b) i) In what stage of its life cycle is the Sun now?
   ii) What will be the final stage of the Sun's life cycle?

7 What process provides the energy for a star to emit its own light?

8 **a)** What is meant by the term main sequence star?
   b) What are the two forces which act on a main sequence star to keep it stable?

9 Why does a white dwarf eventually becomes a black dwarf?

10 **a)** Explain why a red super giant collapses to produce a supernova.
   b) What may happen to the core of a star after a supernova explosion?

11 What is the evidence to suggest that the atoms which make our Earth were once part of a giant star?

○ Circular orbits

Although planets and moons move in elliptical orbits, many of the orbits are very nearly circular. In this section we look at the physics of circular orbits. Figure 8.11 shows the Earth in its orbit around the Sun.

At position E_1, the Earth has a velocity v_1 and by the time the Earth reaches position E_2 it has a velocity v_2. Because velocity is a vector

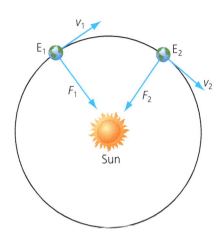

▲ **Figure 8.11** The circular orbit of the Earth around the Sun.

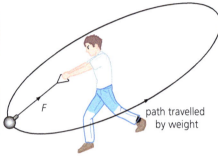

▲ **Figure 8.12** As the hammer turns in a circle, it accelerates.

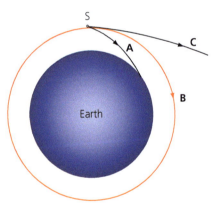

▲ **Figure 8.13** Speed C > speed B > speed A.

quantity, this means that the Earth has changed its velocity in going from position E_1 to E_2. However, the speed of the Earth in its orbit remains constant.

In Chapter 5, you learnt that when an object's velocity changes, it must be accelerating.

Since the Earth's velocity changes during its orbit, it must always be accelerating. It is the pull of gravity which causes the Earth to accelerate. In Figure 8.11 the pull of gravity is shown as the forces F_1 and F_2. Although the Earth is always accelerating towards the Sun, it does not get any closer. This acceleration causes the Earth to change direction. This idea is rather like the hammer thrower in Figure 8.12. Before he throws the hammer he whirls it round in a circle. He applies a large force to change its direction.

In a circular orbit:

- a planet's velocity changes, but its speed remains constant
- the planet accelerates towards the Sun which causes it to change direction
- gravity provides the force to accelerate the planet.

Speed of orbit

Figure 8.13 shows the importance of the speed of orbit. S is a satellite travelling around the Earth.

- If its speed is too great it follows path C and disappears into outer space.
- If its speed is too slow it follows path A and falls to Earth.
- When the satellite moves at the right speed, it follows path B and stays in a stable circular orbit.

The speed of orbit of a moon, planet or satellite can be calculated by first calculating the distance the object moves in one orbit.

$$\text{distance} = 2 \times \pi \times \text{orbital radius}$$

The average speed is then calculated using the familiar equation:

$$\text{average speed} = \frac{\text{distance}}{\text{time}}$$

where time is how long it takes the object to complete one orbit.

Orbits near and far

If a planet is close to the Sun, then the pull of gravity from the Sun is strong. This causes the speed of the orbit to be high. For a planet further away, the Sun's gravity pulls less strongly and the planet moves more slowly.

We can use the data from the table on page 249 to calculate the orbital speeds for Earth and Mercury.

$$\text{orbital speed of Mercury} = \frac{2 \times \pi \times r}{T}$$

$$= \frac{2 \times \pi \times 58 \text{ million km}}{88 \text{ days}}$$

$$= 4.1 \text{ million km/day}$$

$$\text{orbital speed of Earth} = \frac{2 \times \pi \times r}{T}$$

$$= \frac{2 \times \pi \times 150 \text{ million km}}{365 \text{ days}}$$

$$= 2.6 \text{ million km/day}$$

Test yourself

12 **a)** What is the difference between velocity and speed?

   b) A planet is in a circular orbit around a star. Explain why the planet's velocity changes although its speed is constant.

13 **a)** Explain why a planet is accelerating as it goes round the Sun.

   b) What is the direction of this acceleration?

   c) i) When an object is accelerating, explain why an unbalanced force must act on it.

      ii) In the case of the planet, what supplies the force to accelerate it?

14 Saturn has 62 moons. The time of orbit of a moon depends on its distance from Saturn.

The table below shows radius of orbit of some of the moons of Saturn, and the time of orbit.

Moon	Radius of orbit in 1000 km	Time period of orbit in days
Pandora	141	0.63
Aegaeon	168	0.80
Mimas	185	0.94
Enceladus	238	1.37
Tethys	295	1.89
Dione	377	2.74
Rhea	527	4.52

▲ Figure 8.14

   a) Plot a graph of the radius of orbit (*y*-axis) against the time period (*x*-axis).

   b) An astronomer discovers a new moon with a time of orbit 3.5 days. Use the graph to find the radius of the moon's orbit. Show your working on the graph.

15 A comet is a large lump of rock and ice in orbit around the Sun. Figure 8.15 shows the very elongated orbit of a comet.

   a) At what point in the orbit, A, B, C or D, will the comet be travelling

      i) the fastest

      ii) the slowest?

   b) What causes the comet to change speed during its orbit?

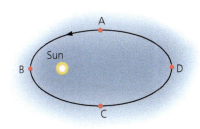

▲ Figure 8.15

Show you can...

Complete this task to show you understand the nature of circular orbits.

The pull of the Earth's gravity makes the Moon accelerate towards the Earth. However, the Moon does not get any closer to the Earth. Explain why.

○ Red-shift

We have all been in a car and seen an ambulance coming towards us and then speed past. As the ambulance goes past us and moves away, we hear the pitch of the sound from the siren become lower (fewer waves per second). The frequency of the sound waves we hear has decreased and the wavelength increased.

A similar effect happens with light waves, but it is only noticeable if the wave source is moving very quickly. When a source of light is moving away from us, the wavelength of the light that we see becomes longer. So the light moves towards the red end of the spectrum. We say the light has been **red-shifted**. If the source of the light moves even faster, the wavelength increases even more producing a greater red-shift.

In 1929 Edwin Hubble began to study the light arriving from distant galaxies. He discovered three important facts about the light from distant galaxies:

- The light emitted from distant galaxies is red-shifted – the wavelength is longer than expected. This tells us that distant galaxies are moving away from us.
- Galaxies appear to be moving away from us in all directions.
- The further away from us a galaxy is, the bigger the red-shift. This tells us that the further away a galaxy is, the faster it is moving.

▲ Figure 8.16 A teacher uses a whirling loudspeaker to model the effect of red-shift. As the loudspeaker moves away from the students, they hear a decrease in pitch. A decrease in pitch means an increase in wavelength. So the loudspeaker moving away from the students produces the same effect as a light source moving away from us.

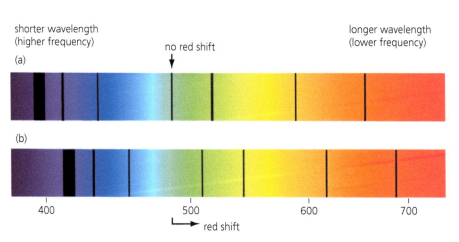

▲ Figure 8.17 a) This image shows light which is emitted from the Sun. It is crossed by black lines. b) This second image shows light arriving at Earth from a distant galaxy. It has the same pattern of black lines as the Sun's light. However, the pattern has been shifted towards the red end of the spectrum. This is how Hubble first discovered the red-shift in the light of distant galaxies.

Big bang theory

Hubble's work on the red-shift of distant galaxies helped to support the **big bang theory**. This theory suggests that the Universe began about 13.8 billion years ago, when all matter and space expanded violently from a single point. Such was the energy of the explosion that the Universe has been expanding ever since. As time passed the hot matter condensed into the atoms we can see today. The force of gravity pulled clouds of gas together to form stars and galaxies.

Figure 8.18 helps you to understand how scientists formed this view.

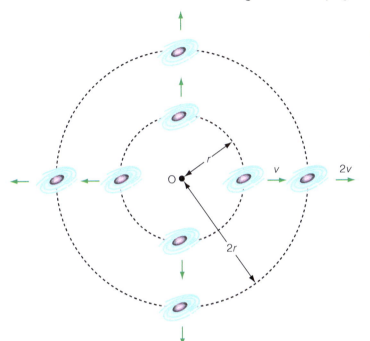

In this diagram you can see galaxies moving away from our Solar System at the centre of the diagram. The galaxies further away move away from us faster. This suggests that at one instant a long time in the past, all the galaxies were in the same position.

Up until the 1990s scientists were confident that the big bang theory explained the origin of the Universe completely. However, recent observations of supernovae have led to the controversial suggestion that the Universe is expanding faster as time passes. This result is a surprise because it had been thought that the effect of gravity would slow down the Universe's expansion.

One explanation is that there is dark energy which is responsible for the Universe's increasing rate of expansion.

▲ Figure 8.18

Test yourself

16 Use the correct answer from the box to copy and complete the sentence.

| go down | not change | go up |

The pitch of a fire engine's siren will _____ as the fire engine drives past you.

17 An astronomer discovers that the light from a galaxy has been shifted to the blue end of the spectrum. What can you deduce from this discovery?
   A The galaxy is moving away from the Earth.
   B The galaxy is not moving.
   C The galaxy is moving towards the Earth.

18 Which of the following statements about the big bang theory are true and which are false? Copy out the true statements.
   A It has been proved correct by mathematical calculations.
   B It is supported by the fact that most distant galaxies are moving away from the Earth.
   C It is based on scientific and religious facts.
   D It is the most satisfactory explanation of present scientific knowledge.
   E It is the only way to explain the origin of the Universe.

19 The table shows the relationship between the speed of a galaxy and its distance away from us.

Galaxy	Distance of galaxy from Earth in 10^{12} km	Speed of galaxy in km/s
A	680	1200
B	3800	6700
C	8500	15000
D	11400	20000
E	22700	40000
F	34000	60000

a) Plot a graph of the speed of the galaxies (*y*-axis) against the distance of the galaxy (*x*-axis).

b) Does this graph help to support the big bang theory? Explain your answer.

c) Two galaxies are discovered to be moving away from us. G is moving with a speed of 2000 km/s. H is moving with a speed of 4000 km/s. Use your graph to work out the distance of each galaxy from the Earth.

20 Explain whether the big bang theory has ever been completely proved.

Show you can...

Complete this task to show you understand the evidence for the big bang theory.

Explain how the discovery of the red-shift of the light emitted from distant galaxies supports the big bang theory.

TIP

The purpose of this section is to provide an example of how new scientific knowledge leads to a change in a theory.

○ Sun at the centre – understanding models

Everyone knows that the Earth goes around the Sun, and that the time taken for the Earth to complete one orbit is a year. However, for thousands of years our ancestors thought that the Sun went around the Earth. It is easy to understand why people thought that. We see the Sun rise and set each day. People often say things such as 'the Sun has gone below the horizon'. Such language suggests the Sun is moving. With the benefit of scientific knowledge, we know that the Sun sets each day because the Earth spins on its axis once every day.

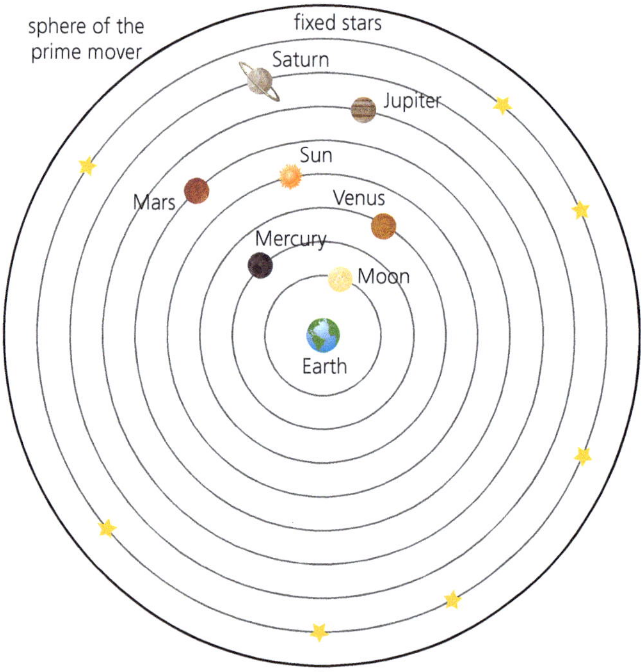

Figure 8.19 shows how Aristotle, the Ancient Greek philosopher, thought the Universe was arranged. He placed the Earth at the centre, and all other heavenly bodies were thought to move around the Earth. He thought that the planets and the Moon moved at different speeds to the stars and that this explained why they appear in different positions relative to the stars.

Copernicus (1473–1543)

So what was it that caused people to change their mind about the Sun's motion? The problem with Aristotle's model of the Universe was that it did not really explain properly the motion of the planets, which appear to wander backwards and forwards across the sky. Figure 8.20 shows how Mars appears to move across the sky over the period of 1 year.

▲ **Figure 8.19** An early idea of the arrangement of the Universe.

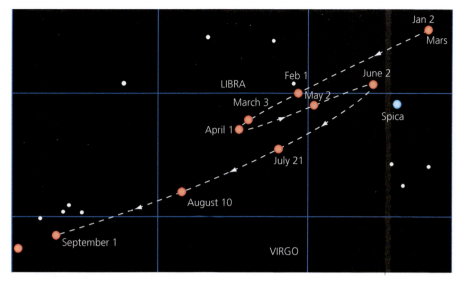

▲ **Figure 8.20** The annual motion of Mars through the stars.

Copernicus realised that the motion of the planets could be explained if the Sun stands still at the centre of the Solar System, and the Earth and other planets orbit around it.

We can understand the backwards loop of Mars when we realise that planets closer to the Sun orbit faster than the more distant ones. Mars orbits around the Sun more slowly than the Earth. About once a year we overtake Mars, so it appears to go backwards as we overtake it.

Galileo

In 1610 Galileo heard about the invention of the telescope and made one for himself. He looked at the Moon, Venus, Mars, Jupiter and Saturn. Galileo made these discoveries.

1 Mars varies in brightness.

2 Venus has phases like the Moon.

3 Jupiter had four moons (visible through a small telescope).

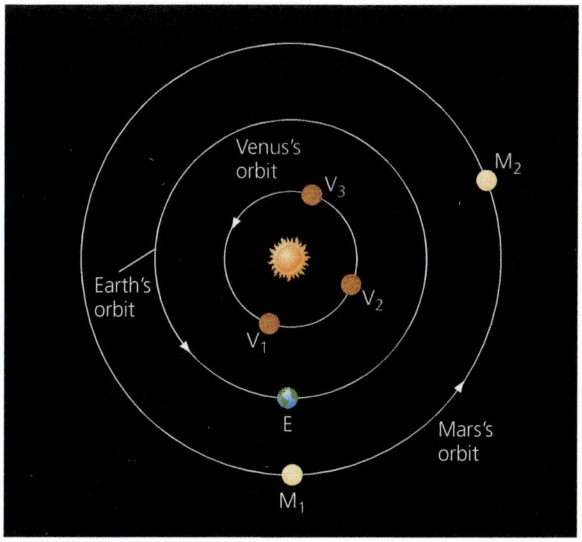

Each of the last points helped to support Copernicus's view of our Solar System. Figure 8.21 helps you understand some of these ideas.

- The distance between Earth and Mars changes so the brightness of Mars changes.
- Venus lies between us and the Sun. Sometimes we can see all of its surface illuminated by the Sun. On other occasions we can only see half of it, or a crescent like a new Moon.

When Galileo looked at Jupiter he could see four moons, which rotated around the planet. On some nights he could only see three moons, because one had gone behind the planet. Galileo could see moons moving round Jupiter, so he thought that the Earth might be moving too.

▲ **Figure 8.21**

▲ Figure 8.22

Test yourself

21 How did Aristotle's view of the Solar System differ from ours?

22 Why did Copernicus think the Earth moved round the Sun?

23 Use Figure 8.21 to help you explain why Mars appears to change in brightness. In which position will it appear brighter M_1 or M_2?

24 Figure 8.22 shows what Galileo saw through his telescope on two different nights.

  a) What are the small dots?

  b) Why do the dots move?

  c) Why did Galileo see three dots on one night and four dots on another night?

  d) Do some research to find the names of the four Galilean moons.

25 Make sketches to show the shape and relative size of venus in position V_1, V_2 and V_3, when viewed through a telescope.

Show you can...

Show you understand the evidence for our model of the Solar System by completing this task.

Summarise the evidence that made us change our model of the solar system from one in which the Sun goes around the Earth, to one in which the Earth goes round the Sun.

Chapter review questions

1 The stages of development of a star much larger than the Sun are listed below. Write down the stages in the correct order.

   A Protostar   **D** Supernova

   B The main sequence   **E** Red giant

   C Dust and gas   **F** Neutron star or black hole

2 a) What force acts on dust and gas in space?

   b) What is the name of the process which releases energy in the core of the Sun?

   c) Why does the Sun not collapse due to the pull of gravity on it?

   d) Jupiter and Saturn are giant planets made mostly out of hydrogen. Why do they not produce their own light as the Sun does?

3 Mercury is a rocky planet, Haumea is a dwarf planet and Gaspra is an asteroid. Gaspra is a long piece of rock with an irregular shape.

   a) i)   What is different between the shapes of Gaspra and Haumea?

      ii) Why are some planets called dwarf planets?

      iii) State one major difference between Mercury and Haumea.

   b) Mercury and Haumea are in motion. What is similar about the motion of the two objects?

4 Explain the nature of each of these types of objects.

   a) Main sequence star   d) Supernova

   b) White dwarf   e) Black hole

   c) Red giant   f) Galaxy

5 The light from most distant galaxies is 'red-shifted'. What does this tell you about the motion of the galaxies?

6 There are over 300 000 known objects in orbit around the Sun.

   The table shows the distances from the Sun and the time periods for the orbits of some of these objects.

   a) Plot a graph of time period of the orbits against the radius of the orbits.

Object orbiting the Sun	Radius of orbit in 10^9 km	Time of orbit in years
Chiron	2.1	51
Bienor	2.6	68
Uranus	2.9	84
Neptune	4.8	165
Pluto	5.9	248
Makemake	6.9	310
Eris	10	557

   b) An astronomer has made a mistake in recording the radius of orbit for one of the objects. Which one? Suggest what the radius of the orbit should be.

   c) Haumea is a dwarf planet with a radius of orbit 6.4×10^9 km. Use the graph to predict its time period of orbit.

Practice questions

1 The sentences below describe the life cycle of a star which is much more massive than the Sun. The sentences are not in the correct order.

    A It expands to form a red super giant.

    B The force of gravity pulls gas and dust together.

    C It is in a stable state as a main sequence star.

    D The core collapses to make a neutron star.

    E It explodes as a supernova.

  a) Copy out the sentences in the correct order. Start with the sentence labelled B. [2 marks]

  b) What balances the pull of gravity in a main sequence star? [1 mark]

2 Stars go through a life cycle. At the end of their life cycle some stars become black holes.

  a) What type of star becomes a black hole at the end of its life cycle? [1 mark]

  b) Describe the nature of a black hole. [2 marks]

3 a) Scientists have measured the wavelengths of the light given out from distant galaxies. The wavelengths are longer than expected.

    i) What name is given to the effect causing the increase in wavelength? [1 mark]

    ii) Use the correct answer from the box to complete the sentence. Copy out the completed sentence.

      | changing colour   expanding   getting brighter |

    The increase in the wavelength of the light from distant galaxies provides evidence to support the idea that the Universe is _____. [1 mark]

  b) Look at the sketch graph.

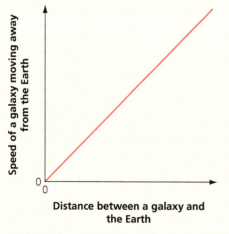

▲ **Figure 8.23**

What pattern links the distance between a galaxy and

the Earth and the speed that the galaxy moves away from the Earth? [1 mark]

  c) There are several different theories about how the Universe was created.

    i) One of the theories suggests that the Universe began at a single point and then expanded violently outwards.

    What name is given to this theory? [1 mark]

    ii) Suggest why scientists support a particular theory. [1 mark]

4 Explain why scientists think that that the Sun and our Solar System were formed out of the remains of a supernova. [2 marks]

5 Describe what these objects are:

  a) A main sequence star [2 marks]

  b) The Milky Way [2 marks]

  c) A supernova [2 marks]

  d) A neutron star [2 marks]

6 Stars go through a life cycle of change.

  a) Explain how a star is formed. [4 marks]

  b) Why does the amount of hydrogen in a star decrease as time goes on? [1 mark]

  c) Describe what happens to a star much larger than the Sun from the time it runs out of hydrogen. [5 marks]

7 Our Solar System contains atoms of the heaviest elements.

  a) Where were these elements formed? [1 mark]

  b) Explain how these elements are formed. [2 marks]

  c) What does this tell you about the age of our Solar System compared with the age of most of the stars in the Universe? [1 mark]

8 a) Light which reaches us from distant galaxies is shifted towards the red end of the spectrum. Explain why. [3 marks]

  b) Explain why the red-shift in the light from distant galaxies provides evidence to support the big bang theory for the origin of the Universe. [4 marks]

9 A satellite is in a circular orbit around the Earth.

  a) The speed of the satellite remains constant. Explain why the velocity of the satellite changes during the orbit. [2 marks]

  b) Explain why the satellite is accelerating constantly while it is in orbit around the Earth. [2 marks]

  c) In which direction is the acceleration of the satellite, and what force is responsible for this acceleration? [2 marks]

Working scientifically

Changing theories, changing models

Our scientific understanding of how the Universe was created has changed and what we now think may still not be correct. Scientists can look at evidence and make suggestions, even change theories, but any theory is only as good as the evidence that supports it. Our current theory, the big bang theory predicted the existence of a background radiation throughout the Universe but this radiation could not be detected. So the discovery in 1965 of cosmic microwave background radiation (CMBR), a radiation that seems to fill the whole Universe, was significant evidence in support of the big bang theory. Currently it is the only theory that can explain the existence of CMBR. So is the big bang theory correct? Maybe and maybe not. What we do know is that it is a good idea that explains some important observations.

In this chapter you have also read how our understanding of the Solar System has changed since the model put forward by the Ancient Greek philosopher Aristotle. The model of the Solar System has changed several times since then. New observations, maybe as the result of new or improved technology, lead to theories, ideas and models being changed or replaced.

Scientists do know that a satellite is kept in its orbit around the Earth by the force of gravity between the satellite and the Earth. The gravitational field model is well established.

Mia used a rubber bung whirling around on a piece of string to model the circular orbit of a satellite around the Earth. The force keeping the rubber bung moving in a circle is the tension in the string. This is caused by the weight of the masses hanging from the string.

Using the same rubber bung and keeping the tension force constant, Mia investigated the effect of changing the speed of the rubber bung on the radius of the circle.

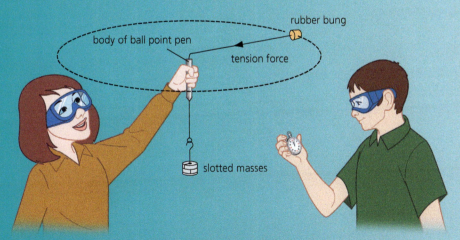

▲ Figure 8.24 In this investigation, the rubber bung is like the satellite and the tension in the string acts like the gravity pull on a satellite.

1 Why was it important to use the same rubber bung and tension force?

While Mia whirled the rubber bung around trying to keep a constant speed, a second student, Chris, timed how long it took the rubber bung to complete one rotation. Mia told Chris that it would be better if he timed 10 rotations rather than just one.

2 Why is it better to time 10 rotations rather than just one?

3 Mia and Chris did the investigation outside and away from other students. They also wore plastic safety goggles.

What risk were the students trying to control?

Mia used the average time for one rotation and the radius of the circle to calculate the average speed of the bung. Mia then plotted the graph shown in Figure 8.25.

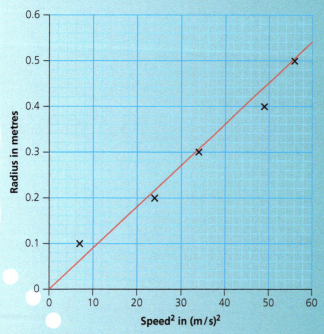

▲ Figure 8.25

4 What should Mia conclude from this investigation about the effect of changing the speed of the rubber bung on the radius of the circle?

A second pair of students did the same investigation. The results these students obtained showed the same pattern as Mia's results.

5 Were the results from the investigation **reproducible**?

6 In this investigation the force on the rubber bung (the tension in the string) did not change as the radius of the circle changes.

How is this different to the gravity pull on satellites orbiting the Earth at different heights?

KEY TERMS

A measurement is **reproducible** if the investigation is repeated by a different person and the same pattern of results is obtained.

9 Cross-chapter Questions

These questions are higher level questions which are designed to test a student's knowledge across two areas of knowledge. For example, question 2 combines calculation about electrical work, with working out specific heat capacity.

1   The driver of a car travelling at 15 m/s uses the brakes to stop quickly. The car's braking distance is 40 m. When the same braking force is used to stop the car from a speed of 30 m/s, the braking distance is significantly more than 80 m. Explain why.   [6 marks]

2   A student uses a heater to warm up the block of metal shown in Figure 9.1. The metal block has a mass of 0.5 kg. After 5 minutes the metal has warmed up from 21°C to 51°C. Calculate the specific heat capacity of the metal.   [6 marks]

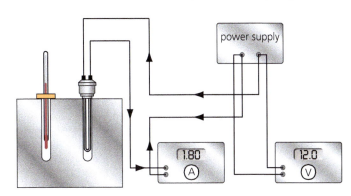

▲ Figure 9.1

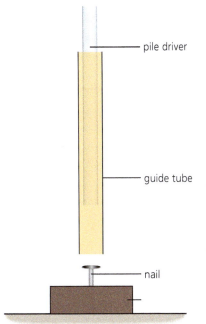

▲ Figure 9.2

3   A student tests a model pile driver by dropping it from a height of 25 cm on to a nail, as shown in Figure 9.2. The nail is knocked into the wood by a further 4 mm. Calculate the average force which acts on the pile driver to bring it to rest. The mass of the pile driver is 0.45 kg.   [6 marks]

4   Explain how the temperature inside a large cold cloud of gas in space can increase sufficiently for a star to be formed from the gas.   [6 marks]

5   A car manufacturer uses exactly the same engine in the two cars, A and B shown in Figure 9.3. Explain why the manufacturer finds that car A uses less petrol than car B when they each travel 100 km in a test drive at a speed of 20 m/s.   [5 marks]

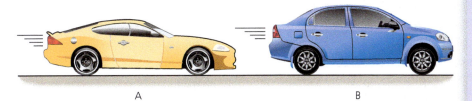

▲ Figure 9.3

6   Explain why ions are formed near a flame and also near a radioactive source which emits alpha particles.   [6 marks]

7   The electric motor shown in Figure 9.4 is used to lift the weight through a distance of 0.8 m in 5 seconds. Show that this process dissipates some energy.   [6 marks]

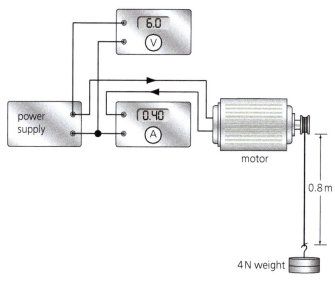

▲ Figure 9.4

8   A girl jumps off a wall and lands on the ground below. She bends her knees when she lands. The girl says that she does not hurt herself by bending her knees because she slows down in a longer time. A boy watching says that she does not hurt herself because she slows down over a longer distance. [6 marks]

Explain why both students are correct.

9   A student inflates a balloon. He then releases the air in the inflated balloon so that the balloon flies around the laboratory. The student says to his friend that he can explain why this happens, using ideas about forces. Another student says that he can explain the balloon's movement using ideas about energy. Explain why both students are correct.   [6 marks]

10 A student suspends a negatively charged ball close to a metal plate. When the plate is charged positively the ball is deflected as shown in Figure 9.5. The student says that the ball's deflection can be explained using ideas about forces. A friend says he can use ideas about energy to explain how far the ball moves. Explain why both pupils are correct.   [6 marks]

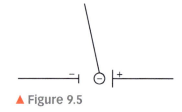

▲ Figure 9.5

11 Figure 9.6 shows a radioactive source which emits beta-particles. A teacher demonstrates to the class that the magnetic field deflects the particles to the left as they pass through the field. Explain why the beta-particles are deflected in this direction.   [4 marks]

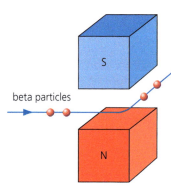

beta particles

▲ Figure 9.6

12 A student drops a small magnet through a solenoid as shown in Figure 9.7. As the magnet enters the coil the student observes that it slows down and a current flows through the ammeter. Explain why both of the effects observed by the student happen.   [5 marks]

13 A convex lens can be used to focus the sun's rays and to burn a piece of paper. Explain why this happens.   [4 marks]

[Safety: you must never look at the sun through a lens, or use the lens to focus the sun's rays on to your skin.]

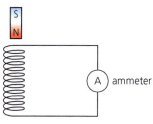

▲ Figure 9.7

14 A meteor is a small piece of ice and rock which orbits the sun. Sometimes the Earth encounters a meteor which is travelling at 20 km/s relative to the Earth. Then the meteor burns up in the atmosphere and it can be seen as a bright streak of light. Explain why the meteor burns up in the atmosphere.   [5 marks]

15 Two boxes A and B have the same volume. When each box is pushed with the same force, B accelerates faster. Explain which box has the higher density.   [5 marks]

Glossary

1 becquerel (1Bq) An emission of 1 particle/second

1 GW, 1 gigawatt = 10^9 W

absorb A medium which does not allow light to pass through it

accurate A measurement or calculated value that is close to the true value

alpha particle Formed from two protons and two neutrons

amplitude The height of the wave measured from the middle (the undisturbed position of the water)

angle of incidence The angle between the incident ray and the normal

angle of reflection The angle between the reflected ray and the normal

angle of refraction The angle between the refracted ray and the normal

anomalous A result that doesn't fit the expected pattern

atomic number Number of protons in an atom

background count The average count rate which a Geiger–Müller tube records over a period of time when the counter is not close to a radioactive source; the background count is caused by radioactive materials in our environment

background radiation Radiation from natural and man-made sources in our environment

beta particle A fast moving electron

black hole The most concentrated state of matter, from which even light cannot escape

braking distance The distance a car travels while the car is stopped by the brakes

calibrate Mark a scale onto a measuring instrument so that you can give a value to a measured quantity

categoric variable A variable with values that are given a name or label

centre of mass The point through which the weight of an object can be taken to act

closed system A system with no external forces acting on it

concave lens Represented by ⟩ in ray diagrams

contact force Can be exerted between two objects when they touch

continuous variable A variable with numerical values obtained by either measuring or counting

control variable The variable which is kept constant

control variable A variable which is kept the same

convex lens Represented by ⟨↑⟩ in ray diagrams

crumple zones Zones at the front and back of cars which are designed to buckle in a crash

dependent variable The variable that changes because of a change made to the independent variable

diffuse reflection When light is reflected at different angles off a rough surface

directly proportional When two quantities are directly proportional, doubling one quantity will cause the other quantity to double; when a graph is plotted, the graph line will be straight and pass through the origin (0, 0)

displacement A distance travelled in a defined direction

dissipate To scatter in all directions or to use wastefully; the energy has spread out and heats up the surroundings

elastic deformation When an object returns to its original length after it has been stretched

electric current A flow of electrical charge; the size of the electric current is the rate at which electrical charge flows round the circuit

electron A negatively charged particle that orbits the nucleus of an atom

ethical issue A question which requires a consideration of what is good or bad for both individuals and society as a whole

extension The difference between the stretched and unstretched lengths of a spring

extrapolation (or prediction) Assuming that an existing trend or pattern continues to apply in an unknown situation

fair test Only the independent variable affects the dependent variable

fluid A liquid or a gas; a fluid flows and can change shape to fill any container

flux density The number of lines of magnetic flux in a given area

focal length The distance between the lens and the principal focus

force A push or a pull

frequency Hz The number of waves produced each second or the number of waves passing a point each second. The unit of frequency is hertz; 1 hertz means there is 1 cycle per second

gamma ray An electromagnetic wave

Geiger-Müller (GM) tube A device which detects ionising radiation; an electronic counter can record the number of particles entering the tube

half-life The time taken for the number of nuclei in a radioactive isotope to halve; in one half-life the activity or count rate of a radioactive sample also halves

hypothesis An idea based on scientific theory; a hypothesis states what is expected to happen

independent variable The variable that you change; an investigation should only have one independent variable

induced magnet A magnet which becomes magnetic when it is placed in a magnetic field

inelastic deformation When an object does not return to its original length after it has been stretched

inertia Inactivity; objects remain in their existing state of motion – at rest or moving with a constant speed in a straight line – unless acted on by an unbalanced force; the inertial mass of an object is a measure of how difficult it is to change its velocity. We can define the inertial mass through the equation

$$m = \frac{F}{a}$$

intensity of radiation The power of the radiation incident per square metre

interval The difference between one value in a set of data and the next

inversely proportional When two quantities are inversely proportional, doubling one quantity will cause the other quantity to halve

isotopes Different forms of a particular element; isotopes have the same number of protons but different numbers of neutrons

limit of proportionality The point beyond which a spring will be permanently deformed

longitudinal wave A wave in which the vibration causing the wave is parallel to the direction of energy transfer

magnetic Magnetic materials are attracted by a magnet

main sequence A star that releases energy by fusing hydrogen to form helium

mass number Number of neutrons plus protons in an atom

moon A natural satellite in orbit around a planet

neutron A neutral particle found in the nucleus of an atom

neutron star A very dense small star made out of neutrons

non-contact force Can sometimes be exerted between two objects that are physically separated

non-ohmic The current flowing through a non-ohmic resistor is not proportional to the potential difference across it; the resistance changes as the current flowing through it changes

non-renewable energy resources Energy resources which will run out and cannot be replenished

normal A line drawn at 90° to a surface where waves are incident

north-seeking pole The end of the magnet that points north

nuclear fusion The name given to the process of joining small nuclei together to form larger ones

ohmic The current flowing through an ohmic conductor is proportional to the potential difference across it; if the p.d. doubles, the current doubles but he resistance stays the same

opaque An object which does not allow light to pass through

peer review A process by which scientists check each other's work

period The time taken to produce one wave

permanent magnet A magnet which produces its own magnetic field; it always has a north pole and a south pole

planet A large body which orbits the Sun

potential difference (p.d.) A measure of the electrical work done by a cell (or other power supply) as charge flows round the circuit; potential difference is measured in volts (V)

$$\text{power} = \frac{\text{energy transferred}}{\text{time}}$$

precise A set of measurements of the same quantity will closely agree with each other

prediction An extrapolation arising from a hypothesis that can be tested; experimental results that agree with the prediction provide evidence to support the hypothesis

principal focus (of a convex (converging) lens) The point through which light rays, parallel to the principal axis, pass after refraction

principle of moments When a system is balanced the sum of the anti-clockwise turning moments equals the sum of the clockwise turning moments

proton A positively charged particle which is found in the nucleus of an atom

protostar A large ball of gas contracting and heating to form a star

P-waves Longitudinal seismic waves

random error An unpredictable error; repeating the measurement and then working out a mean will reduce the effect of a random error; in a graph, data that is scattered about the line of best fit shows a random error

range The maximum and minimum values used or recorded

real image Formed when light rays converge to a point; a real image can be projected onto a screen

red giant A very large star which fuses helium into heavier elements

relay A device which uses a small current to control a much larger current in a different circuit

renewable energy resources Energy resources which will never run out and are (or can be) replenished as they are used

repeatable Measurements are repeatable when the same person repeating the investigation under the same conditions obtains similar results

reproducible A measurement or pattern of results which can be repeated if the same investigation is repeated by a different person

resistor Acts to limit the current in a circuit; when a resistor has a high resistance, the current is low

resolution The smallest change an instrument can detect

resultant force A number of forces acting on an object may be replaced by a single force that has the same effect as all the forces acting together. This single force is called the resultant force.

risk The probability that something unpleasant will happen as the result of doing something

risk assessment An analysis of an investigation to identify any hazards, possible risks associated with each hazard, and any control measures you can take to reduce the risks

scalar A quantity with only magnitude (size) and no direction

slip-rings Rings which provide a continuous contact between the rotating sides of the coil and the connections to the oscilloscope marked ① and ② in Figure 7.34 (p. 234)

south-seeking pole The end of the magnet that points south

specular reflection When light is reflected off a smooth surface in a single direction

stopping distance The sum of the thinking distance and braking distance

supernova A gigantic explosion caused by runaway fusion reactions in a very large star

S-waves Transverse seismic waves

systematic error A consistent error, usually caused by the measuring instruments, when all of the data is higher or lower than the true value; data with a systematic error will give a graph line that is higher or lower than it should be

terminal velocity When the weight of a falling object is balanced by resistive forces

thinking distance The distance a car travels while the driver reacts

translucent An object allows light to pass through, but we cannot see objects through it clearly; some plastics and frosted glass are translucent

transmit A medium which allows the light to pass through it

transparent An object which allows us to see clearly through it; glass is transparent

transverse wave A wave in which the vibration causing the wave is at right angles to the direct of energy transfer

ultrasound A sound wave with a frequency above the range of human hearing (above 20 kHz)

uncertainty For a set of measurements, the difference between the maximum value and the mean or between the minimum value and the mean gives a measure of the uncertainty

valid A method of investigation which produces data that answers the question being asked

vector A quantity with both magnitude (size) and direction

velocity A speed in a defined direction

virtual image An image formed by light rays which appear to diverge from a point

wavelength The distance from a point on one wave to the equivalent point on the next wave

white dwarf A star which is no longer able to fuse helium

work = force × distance

zero error When a measuring instrument gives a reading when the true value is zero

INDEX